TOWARDS AN ISA ENVIRONMENTAL MANAGEMENT STRATEGY FOR THE AREA

ISA Technical Study No: 17

ISA TECHNICAL STUDY SERIES

ISA TECHNICAL STUDY NO: 17

TOWARDS AN ISA ENVIRONMENTAL MANAGEMENT STRATEGY FOR THE AREA

Report of an International Workshop convened by the German Environment Agency (UBA), the German Federal Institute for Geosciences and Natural Resources (BGR) and the Secretariat of the International Seabed Authority (ISA) in Berlin, Germany, 20-24 March 2017

NATIONAL LIBRARY OF JAMAICA CATALOGUING-IN-PUBLICATION DATA

Towards an ISA environmental management strategy for the area:
    report of an international workshop convened by the German
Environment Agency (UBA), the German Federal Institute for Geosciences
and Natural Resources (BGR) and the Secretariat of the International
Seabed Authority (ISA) in Berlin, Germany, 20-24 March 2017.
    p. : col. ill.; cm — (ISA Technical study; 17).
Includes index.
ISBN 978-976-8241-48-1 (pbk)
ISBN 978-976-8241-49-8 (e-bk)

1.   Ocean mining – Environmental aspects
2.   Marine resources conservation – Congresses
3.   Environmental management – Law and legislation
4.   Environmental monitoring
I.   International Seabed Authority   II.   German Environmental Agency (UBA)
III.   German Federal Institute for Geosciences and Natural Resources (BGR)
IV. Series

333.916416   dc 23

International Seabed Authority
14-20 Port Royal Street
Kingston, Jamaica
Tel: +1 876 922-9105
Website: www.isa.org.jm

Table of Contents

1.   Foreword

An International Workshop entitled "Towards an ISA Environmental Management Strategy for the Area" was held in Berlin, Germany, from the 19th to 24th March 2017.

The workshop was jointly organised by the German Environment Agency (UBA) on behalf of the Federal Ministry for the Environment, Nature Conservation, Building and Nuclear Safety (BMUB), the German Federal Institute for Geosciences and Natural Resources (BGR) on behalf of the Federal Ministry for Economic Affairs and Energy (BMWi, and the Secretariat of the International Seabed Authority (ISA), and was supported by the Institute for Advanced Sustainability Studies Potsdam e.V. (IASS).

The primary intention of the workshop was to discuss and to provide a first structured and interdisciplinary input to the Discussion Paper on the development and drafting of Regulations on exploitation for mineral resources in the Area (environmental matters), hereafter referred to as the "Discussion Paper" or the "Draft EnvRegs", which was published by the ISA in January 2017. Furthermore, the workshop aimed at assisting the ISA in developing a long-term Environmental Management Strategy for the Area. Several environmental aspects of a future governance of deep seabed mining were discussed in more detail, such as overarching principles, substantive criteria, the need for effective environmental standards and environmental impact assessments, the potential of adaptive management and a tiered assessment and management approach.

One hundred experts and stakeholders from a wide range of geographical regions, with a broad variety of professional backgrounds and interests ranging from academia, civil society and contractors to competent authorities, (inter) governmental bodies and members of the Legal and Technical Commission (LTC), attended the workshop.

In order to best involve all participants in the discussions and to capture the wide range of opinions present in the room, participatory and interactive working formats were chosen. Thus, following on from introductory plenary presentations of thematic background discussion documents by leading experts, which were distributed to all participants prior to the event, topics were discussed under Chatham House rules either by applying the world café method or in smaller working groups (see Appendix 11.1). The world café method is a structured conversational process in which groups of people discuss a topic or question at several tables, with individuals switching tables periodically and getting introduced to the previous discussion at their new table by a "table host". This method was received very positively by the participants and greatly facilitated conversation between all stakeholders.

This workshop report aims at reflecting as comprehensively as possible, based on written input from presenters, moderators as well as the audio-recordings of the plenary sessions and within the constraints of report length, the discussions, ideas and views of all participants that came out of the workshop.

2. Preface

The workshop, Towards an ISA Environmental Management Strategy for the Area, makes an important contribution to the continued development and delivery of the Authority's Mining Code, under the umbrella of the United Nations Convention on the Law of the Sea, to enable deep seabed mining activities in the Area, while ensuring the effective protection of the marine environment.

The Authority recognises the importance of expert workshops in making an invaluable contribution to its work programme and preparation as a fit-for-purpose regulator of mining activities in the Area. They also provide an important forum for the Authority to engage openly with its expert stakeholder base.

Much work goes into the planning and execution of these workshops, including the drafting of background discussion papers by individual participants and the work of a steering committee. Consequently, I am very appreciative for the exemplary preparation and workshop management undertaken by our co-hosts, the German Environment Agency and the German Federal Institute for Geosciences and Natural Resources and to all participants for their expert contribution and efforts in identifying solutions to, at times, complex issues.

We are not starting from scratch. There are decades of experience we can draw upon, gained in the regulation of related land-based and offshore extractive industries. There is existing exploration activity in the Area focused on data collection and assessment. These activities together with the implementation of the Authority's data management strategy and plan, will allow the Authority and its stakeholders to survey the depth of knowledge we have and to target those areas that need further study and investigation. Improved data gathering and information will allow the Authority to adapt its regulatory framework moving forward and to make better and informed decisions.

Environmental regulation is one of the most important tasks of the Authority under the Convention and one of the priority issues for consideration under the 1994 Agreement. Our continued development of a Mining Code that fosters the sustainable development of mineral resources in the Area is an imperative. Additionally, I am also conscious of a need for the Authority to build a regulatory mechanism that supports invention, innovation and technological development to facilitate the delivery of commercial, economic and environmental objectives and solutions.

Michael W Lodge
Secretary-General
International Seabed Authority

3. Acronyms and Abbreviations

Abbreviation	Name
AM	Adaptive Management
APEI	Areas of Particular Environmental Interest
Authority (the)	International Seabed Authority (also ISA)
BAT	Best Available Technology
CBD	Convention on Biodiversity
CCZ	Clarion-Clipperton Fracture-Zone
CHM	Common Heritage of Mankind
DSM	Deep Sea Mining
DOSI	Deep Ocean Stewardship Initiative
EBSA	Ecologically and Biologically Significant Areas
EIA	Environmental Impact Assessment
EIS	Environmental Impact Statement
EMMP	Environmental Marine Management Plan
EMP	Environmental Management Plan
EnvRegs	Environmental Regulations
ERA	Ecological Risk Assessment
IMO	International Maritime Organisation
IRZ	Impact Reference Zone
ISA	International Seabed Authority (also the Authority)
ITLOS	International Tribunal for the Law of the Sea
LC/LP	London Convention/London Protocol
LTC	Legal and Technical Commission
MPA	Marine Protected Area
NGO	Non-Governmental Organisation
PMT	Pilot Mining Test

PoW	Plan of Work
PRZ	Preservation Reference Zone
REA	Regional Environmental Assessment
REMP	Regional Environmental Management Plan
RRP	Rules, Regulations and Procedures
SDG	Sustainable Development Goal
SEA	Strategic Environmental Assessment
SEMP	Strategic Environmental Management Plan
TRL	Technical Readiness Levels
UNCLOS	UN Convention on the Law of the Sea
VME	Vulnerable Marine Ecosystem

4.   Executive Summary: Points for Further Consideration

The Berlin Workshop was attended by approximately one hundred experts from a wide variety of professional backgrounds and geographical regions, and had two main objectives:

1. To provide feedback on the Discussion Paper on the development and drafting of Regulations on Exploitation for Mineral Resources in the Area (Environmental Matters) (Draft EnvRegs) published by the ISA in January 2017, and to deliver stakeholder-based input to the future Environmental Regulations on exploitation for mineral resources in the Area currently under development; and

2. To discuss the scope for a long-term and overarching ISA Environmental Planning and Management Strategy for the Area.

The workshop organisers, in cooperation with the Steering Committee, put forward the following "points for further consideration". These were presented to and discussed with the participants during the final workshop session and take account of various comments and suggestions received in writing after the workshop. These "points for further consideration" reflect both the breadth of the discussions and the input received during the workshop. However, they by no means reflect a consensus between participants on specific topics, unless indicated otherwise. Any diverging views and opinions have been captured as completely as possible in the following chapters of this report.

General Comments on the Discussion Paper

1. In general, the Discussion Paper was seen as a good starting point for discussion; however, many proposals for improvements to the Draft EnvRegs were raised, including considerations of the level of prescription in the regulations versus guidelines, recommendations, etc.

2. The overall proposed structure of the Draft EnvRegs should be further streamlined in order to fulfil the aim of providing a procedure that clearly separates tasks for the applicant, the ISA and the sponsoring State, respectively.

3. The DOSI process flowchart based on the Draft EnvRegs (Figure 1) was considered to be a helpful tool to better understand and to further discuss the Draft EnvRegs. However, it is recognised that as the Draft EnvRegs evolve, the flowchart will also need to evolve.

4. It was considered whether it might be a helpful approach to combine the ISA *Working draft of the regulations and standard contract terms on exploitation for mineral resources in the Area* (2016), the Draft EnvRegs and the upcoming Draft Seabed Mining Inspectorate Regulations into one final document.

5. It was recommended to further build on and make more efficient use of other competent international organisations' existing body of work.

6. The section "Use of terms and scope" needs further work and precision in defining terms.

Substantive Criteria

7. Environmental objectives will be crucial in order to determine the acceptable level of "harmful effects", consistent with Article 145 of UNCLOS and therefore need further elaboration:

   - Environmental objectives could be general for all mineral resources. However, the translation into acceptable levels of "harmful effects" probably requires a separate approach for each resource category;

   - It was considered essential to develop acceptable impact criteria for the application process (evaluation criteria for use by the LTC);

   - It was proposed that seabed integrity be used as one potential parameter for the assessment of effects of mining activities on the seafloor environment, complemented by others such as species richness, community structure and ecosystem functions, whilst not excluding much needed research on pelagic systems/water column impacts associated with return water discharges and sediment plumes;

   - A "marine environmental health index" composed of eight indicator variables was proposed as an option for defining a "good status" of the seabed;

   - The long-term preservation of sufficiently large, ecologically representative and connected areas was proposed as a key environmental management tool for achieving effective protection of the marine environment.

8. Without prejudice to the terms of the Convention and the 1994 Agreement, criteria for the approval of a Plan of Work (PoW) require elaboration. Such criteria could include whether an applicant has been able to demonstrate, or has taken all reasonable steps to demonstrate, its ability to put in place an effective management system "to ensure effective protection for the marine environment from harmful effects which may arise" from exploitation activities. Furthermore, an application may not be successful if there is insufficient data (needs definition) to assess the effects of the proposed PoW on the marine environment, or insufficient monitoring capabilities.

Roles and Responsibilities

9. A need was seen for clarification of the division of responsibilities and tasks among sponsoring States, the ISA and its organs in order to allow for effective supervision and enforcement of contractor activities in the Area.

10. The ISA must have the capacity to effectively control and assess the activities of contractors in a timely manner, and to ensure that the rules are effectively enforced.

11. Clarification is necessary with regard to which organ of the ISA should be responsible for the various actions to be taken, and where cost-effective, possibly including consideration of a new organ or a new section within the Secretariat responsible for environmental matters.

12. There are matters of jurisdictional competence for "activities in the Area" that require clarification i.e. the role of the ISA, States Parties, sponsoring States, flag States, etc. This will need to be reflected in further drafts of the Exploitation Regulations.

General Principles and approaches

13. The three pillars of transparency - access to environmental information, public participation and access to justice - were seen to be essential. Further consideration is needed on how to "operationalise" all three pillars, including access to justice in the context of the common heritage of mankind.

14. The definition of 'Interested persons' in the Draft EnvRegs was discussed as being too narrow as it is limited to "directly affected" persons "in the opinion of the Authority". The definition of "interested persons" and the stakeholder engagement process should match the standards of other international frameworks.

15. Application of the precautionary approach was considered in the ISA Discussion Paper. Further thought needs to be given to how the precautionary approach can be embedded in the regulatory framework and structure.

16. The operationalisation of the ecosystem approach needs further consideration. Currently, the ability of science to define and measure ecosystem-related parameters and functions is limited in the deep sea realm.

Environmental Standards

17. A critical need, while building on existing international standards, is the development of ISA environmental standards (area/resource specific) for various processes and topics, including e.g. a framework for risk assessment, the determination of environmental thresholds and trigger points on the basis of suitable state and pressure indicators, reporting and monitoring, and others. An integrated, multi-stakeholder process for the development of environmental standards is seen as an effective and reasonable approach. Environmental standards should have scientific considerations as their primary basis.

18. Monitoring and reporting on performance standards as defined in the above process, should generally be compulsory. One benefit would be that a level playing field could be created. Some flexibility on the ways and means to achieve the required performance (desired outcome) could be appropriate.

Environmental Impact Assessment

19. Environmental Impact Assessment (EIA) is without doubt an important tool and its content, roles and functions should be clearly specified.

20. The specific requirements and procedures for the overall EIA process should be clearly formulated, including which body undertakes different processes. EIAs should be publicly available for review and comment, as part of the evaluation procedures for approval of an EIA by the ISA. It was proposed that the EIAs should be independently reviewed by scientific experts.

21. Baseline survey standards as specified for the exploration phase, will need to be updated for the purposes of exploitation to reflect more comprehensive spatial and temporal requirements and other measurements to underpin effective EIAs.

22. Potential effects beyond "the Area" must be considered in the EIA, and adjacent coastal states be consulted.

23. The Environmental Impact Statement (EIS) template is being redrafted. Guidelines to support content definition of the EIS should be developed further.

24. Scoping was proposed to be a mandatory step. Through scoping, the aspects to be addressed in the EIA should be determined.

25. Information gathered through an EIA should feed into regional management mechanisms and vice versa.

26. Environmental Risk Assessment (ERA) should be an integral part of the EIA process.

Adaptive Management

27. Adaptive Management (AM) was seen to be crucial in order to proceed in situations where there is uncertainty that cannot be resolved before development, as well as to ensure a precautionary reaction to unanticipated effects.

28. AM should be seen as a tool for environmental risk management of specific projects.

29. AM requires a cautious and gradual development and application of equipment and collection systems to a certain extent in order to allow for adaptive measures.

30. At the project level, the implementation of risk management is principally a contractor responsibility based on any applicable ISA guidelines.

31. AM should not be used as a substitute for binding regulations necessary to protect the environment and avoid harmful effects. The overall question is whether any adaptive management regime can be considered consistent with a precautionary approach, which takes into account the extent to which it can reduce uncertainty and risk.

32. There is a potential for AM to impact security of tenure if overly prescriptive.

33. AM, when used, should be considered as part of the Environmental Management and Monitoring Plan (EMMP). Active AM rather than passive AM should be required.

34. Use of the periodic review process of the PoW (or the individual EMMP) was considered as an additional tool (to active AM by the EMMP) through which to consider new knowledge, information and experiences. The review mechanism should include recommendations as to possible adjustments to necessary measures to secure effective protection of the marine environment or to prevent serious harm thereto. The extent to which any recommendation should be mandatory requires clarification.

35. The EMMP must contain measureable thresholds at which pre-agreed AM responses can be triggered. Nearing or exceeding these thresholds may lead to compliance notices/warnings or to specific actions being specified by the ISA.

36. Effective monitoring of activities by the ISA, including the capacity for 'real time' monitoring and assessment and notification when thresholds are being approached, and mandating actions (trigger points) to be taken where necessary to avoid exceeding thresholds, were seen as being crucial functions of the ISA (the Inspectorate).

Test Mining

37. Testing of collecting systems and equipment was seen as an important step.

38. It is necessary to clarify the role of testing in the overall procedures.

39. The decision and level of testing is primarily a commercial decision.

40. The type of equipment and technological solutions used to optimise the environmental performance of deep seabed mining are crucial for determining and minimising impacts on the marine environment. Thus, testing of equipment and collection systems with regard to their environmental impacts, including verification of modelling results (e.g. for plumes), is seen as highly important.

41. The type of technology and its environmental performance is extremely important for environmental protection. The iterative definition and achievement of Best Available Technology (BAT) is important. Test mining could and should provide information to this end.

Tiered Governance Approach

42. There was support for a tiered approach to ocean environmental management, including environmental objectives and data collection, from an overarching and strategic scale, through the regional level and down to the project-specific level. Specifically, it was suggested that an overarching strategic policy document (high-level SEA-SEMP or "environmental strategy") and individual regional or (sub-regional) management plans (REMPs) could be useful:

- There is a need for a transparent, inclusive and accountable process;

- Planning mechanisms should consider, where applicable, cumulative effects, multi-sectoral uses and alternatives (location, technique and conceptual) in accordance with the Convention;

- Planning mechanisms should be tied to project approvals;

- Prime responsibility is with the ISA but, where practical, in cooperation with other competent international organisations, contractors and independent researchers, as appropriate. The role of the sponsoring State, if any, has to be defined;

- Where practical, mutually beneficial collaboration or cooperation with other competent international organisations or institutions, such as the IMO and UNEP, including, where appropriate, regional sea conventions institutions, will be required in order to give reasonable regard to other legitimate users of the marine environment;

- It was suggested by some participants to include provisions on planning mechanisms in the Draft EnvRegs but others recommended to take such provisions out;

- Spatial management is seen to be crucial. Regional environmental management plans should be in place before EIAs are carried out but to do this funding mechanisms and the commitment of States Parties are required.

43. The CCZ EMP is a good first step in guiding the development of regional management plans:

- The definition and determination of APEIs or similar protected areas should primarily be based on scientific criteria, in particular on ecological representativeness and other more comprehensive criteria;

- Additional data is required by the ISA to further enhance or revise the EMP and inform further decision-making;

- The role of Impact Reference Zones (IRZ) and Preservation Reference Zones (PRZ) can be important but clear management objectives as well as technical criteria for their design need to be developed further (workshop proposed);

- Monitoring is necessary for future decision-making. Thus, this has to be organised and funded;

- REMPs should be reviewed and updated periodically on the basis of new scientific information or analyses. This, in turn, may require modifications to project-specific EMMPs.

Science

44. Identification of gaps in science: there is a need to identify gaps and to target research at appropriate scales, which may require several nations working together. It would be helpful if scientific efforts not only focuses on the basic research aspect but also integrates environmental management topics in relation to mining activities. Funding mechanisms for large-scale, coordinated international research programs have to be clarified and initiated.

Next Steps

Possible workshops (under ISA consideration or otherwise) on the following topics were proposed:

- Spatial planning, including IRZ/PRZs;

- Environmental Standards;

- CCZ-EMP update, including APEIs, as well as EMPs for other regions;

- Workshop to identify thresholds and gaps in scientific knowledge, and to target research;

- EIA in practice (what can we learn from EIA practitioners), including mechanisms for adaptive management;

- Workshop on duties, obligations and institutional arrangements between the ISA, sponsoring States and contractors.

The International Seabed Authority's Secretariat informed the participants that the Legal and Technical Commission ("the Commission") has requested the ISA's Secretariat to revisit the *"Working draft of the regulations and standard contract terms on exploitation for mineral resources in the Area"* issued in July 2016 before the Commission's forthcoming meeting beginning 31 July 2017, taking into account the outcomes of this and other workshops. The Commission will likely deliver a further progress report to the Council at the twenty-third Session, including a proposed road map and timeline for completing the Mining Code. It is anticipated that the Commission will issue a formal request for stakeholder comments on a draft set of Environmental Regulations, possibly as part of a comprehensive first draft of the Exploitation Code.

5. Welcoming Remarks and Keynote Presentations at the Opening Session of the Workshop

5.1 Welcoming Remarks by Prof. Dr. R. Watzel, President of the German Federal Institute for Geosciences and Natural Resources (BGR)

In his welcoming remarks on opening the joint UBA/BGR/ISA workshop in Berlin, Prof. Dr. Watzel emphasised the importance of Article 145 of UNCLOS for the effective protection of the marine environment from harmful effects which may arise from activities in the Area, but also stressed that substantial global increases in metal consumption fuelled by rapid technological innovation will continue to increase the interest in deep seabed mining in the future. Full compensation for this rising demand through large-scale metal recycling is and should be our long-term goal, but is presently not possible due to insufficient technologies and/or logistics as well as low profitability. Thus, a large proportion of the metal demand must still be covered by primary raw materials – either from land or from the ocean. Prof. Watzel speculated that the geological availability of such raw materials in land-based deposits will not constitute a problem over the next decades but that the significant market power of producing countries, insecurity of supply, the increasing low grade of land deposits, poor accessibility, and potential for social conflicts complicate consistent recovery.

We still know very little about the deep sea and about the impact human activities are having and could have on the seabed and its ecosystems, posing a great challenge to the drafting of Environmental Regulations for deep seabed mining at this moment in time and highlighting the necessity for a clear, transparent and adaptive regulatory process. Prof. Watzel emphasised Germany's intention to play an active and responsible role in the development of good environmental standards, rules and regulations for deep seabed mining, which is expressed through its active engagement in deep-sea research and ultimately also in the organisation of this workshop. Finally, he wished the ISA and all workshop participants every success in establishing a regulatory framework for deep seabed mining in a way that reasonably considers all of the interests of human society: its raw material needs, the equitable sharing of benefits, and the effective protection of the marine environment.

5.2 Welcoming Remarks by Dr. Lilian Busse, Head of the Division Environmental Health and Protection of Ecosystems of the German Environment Agency (UBA)

In her opening remarks, Dr. Lilian Busse particularly addressed the environmental perspective of deep seabed mining. She noted that the current development of guidelines and regulations for deep seabed mining has to be regarded in the framework of current economic and political conditions. Having been declared as the Common Heritage of Mankind in UNCLOS, the economic expectations with regard to the Area and its resources were enormous, some even considering deep seabed mining to be the solution to the eradication of poverty on Earth. Over the last decade, it seems that these assumptions need to be corrected. Moreover, other uses of the deep sea need to be given more consideration. The political opinion seems very clear: The UN global Sustainable Development Goals (SDGs) clearly commit society to a sustainable work programme. SDG 14 outlines the following: "Conserve and sustainably use the oceans, seas and marine resources for sustainable development". Dr. Busse also stressed that the G7 explicitly mentions application of the precautionary approach, involving relevant stakeholders and supporting legislation and environmental impact assessment as well as scientific research. Dr. Busse wondered how these economic, political and environmental expectations could be brought together. Several suggestions were made: the precautionary approach should be applied; there is a need to think long-term and for regional and overarching planning instruments and an independent monitoring program has to be designed as a basis to deal with the existing uncertainties. Dr. Busse expressed the hope that the Berlin workshop would allow progress in this regard and thanked the workshop organisers, the steering committee, and the speakers and authors of the background papers for their input.

5.3   Welcoming Remarks by Mr. Michael Lodge, Secretary-General of the International Seabed Authority

In his opening remarks, the Secretary-General acknowledged the work of the German Environment Agency and the German Federal Institute for Geosciences and Natural Resources in the planning and preparation of the workshop and its timely contribution to the Discussion Paper issued by the ISA Secretariat in January 2017. Mr. Lodge continued by emphasising three points.

Point 1: We should be mindful of the provisions of the Convention and the precise nature of the legal mandate given to the Authority: Mr. Lodge underscored the precise nature of the Authority's mandate flowing from Article 145 of the Convention. As a very precisely worded provision, it complements the general provisions of Part XII of the Convention. He stressed that the Authority has precisely defined and limited powers and functions under the Convention, with respect to exploration and exploitation activities in the Area. Its task is to set the conditions under which mining can proceed without causing serious harm to the marine environment. That means preventing, reducing and controlling known significant harmful effects as far as possible through appropriate risk assessment, long-term monitoring and management of environmental effects, and the incentivisation of engineering and mining planning solutions that minimise environmental damage. It does not mean eliminating all harm to the marine environment.

Point 2: We should not reinvent the wheel: Recognising there are no direct comparisons with seabed mining activities, some activities may be similar to other extractive industries. Consequently, in the setting of standards and best practice we should draw on good industry and environmental practice already existing in land-based mining and offshore oil and gas operations. Any gaps can be bridged by developing specific technical standards based on sound scientific evidence, taking account of appropriate technical and economic constraints. Mr. Lodge also noted that we are not starting from scratch and underlined the importance of the existing exploration framework for gathering data, conducting baseline studies and carrying out equipment testing. These activities would continue through the exploitation phase, leading to the development of new and improved technologies.

Point 3: We should be mindful of the scale of activities to be regulated: Mr. Lodge noted that although the Authority is regulating a new activity, a rational and incremental approach needs to be taken. The industry is not going to happen overnight and contractors will advance their activities according to different timescales. He detailed a series of blatant misstatements in recent headlines such as 'invisible land grab' together with misleading comparisons to disasters such as the Deepwater Horizon incident. Mr. Lodge called for a realistic assessment of the likely scale of effects from deep seabed mining in the first 15 years of commercial activity, in order to consider the number of operations and the actual spatial and temporal effect of mining at project level and at regional level and what disasters may happen at sea during a mining operation. Mr. Lodge considered it more useful to focus discussions on what is an acceptable environmental effect, rather than devote time to defining abstract terms, such as 'serious harm'. In isolation, such a term means different things to different people and has different meanings in different contexts.

Mr. Lodge suggested a more valuable exercise would be to look for a threshold based on combining the probability of the occurrence of an incident with the magnitude of its injurious impact. The risk that flows from an activity is primarily a function of the particular application, the specific context and the manner of operation. He noted that the International Law Commission had looked at the issue of legal thresholds, such as 'significant' or 'serious', finding them necessarily ambiguous, with a determination to be made on a case-by-case basis involving more issues of fact than law.

Finally, Mr. Lodge expressed his confidence that collectively, through focused conversations at expert level, the necessary solutions to building a Mining Code that allows for the sustainable development of the mineral resources in the Area can be identified.

5.4 Opening Keynote Presentations

5.4.1 Potential Impacts of Exploitation Activities on the Marine Environment
Antje Boetius and the JPI-O team

Mining regulations need input from deep-sea science

The UNCLOS policy, which prompts avoidance of serious harm to the marine environment and more recent multi-national agreements such as the EU Marine Strategy on 'Good Environmental Status' and the UN Sustainable Development Goal 14 create a strong need for the environmental regulation of deep seabed mining. To date, a total of 27 licenses for the exploration of mineral deposits of the seafloor in areas beyond national jurisdiction ('the Area') have been granted by the ISA. Seven of the licenses for the exploration of manganese nodules in the Clarion-Clipperton Fracture Zone (CCZ) have already expired, providing the licensees with the opportunity to submit an application for exploitation. All seven licensees applied for a five-year prolongation of their exploration contracts, and it thus has to be expected that the ISA will receive applications for exploitation in a few years' time. This puts pressure on the ISA to develop the necessary Environmental Regulations but at the same time urges deep-sea scientists to study the characteristics of the ecosystems associated with mineral deposits and to provide guidance on appropriate indicators and technologies to monitor their status and possible adverse changes. In addition, scientific knowledge can help with the identification of appropriate technologies for the reduction of the environmental impacts of mining equipment.

Recent projects address potential impacts of deep seabed mining

The European multi-national and multi-disciplinary research projects JPI-Oceans "Mining Impact" (http://jpio-miningimpact.geomar.de) and MIDAS (www.midas.eu) study the environmental impacts of deep seabed mining with special emphasis on the exploitation of polymetallic nodules. Several research expeditions were carried out to characterise nodule areas in the tropical Eastern Pacific and to study possible impacts of mining operations. As mining has not yet commenced, local and small-scale seafloor disturbances from sampling activities (e.g. with dredges and other towed gear) and benthic impact experiments were used as analogues of mining operations and compared to undisturbed reference areas outside the disturbance tracks. Some of the disturbances originate from experiments carried out more than 30 years ago. Investigations took place in several license areas, the ISA's Areas of Particular Environmental Interest (APEIs) in the CCZ, and in the largest existing mining-simulation experiment conducted by repeated ploughing of a >10 km^2 area in the DISCOL Experimental Area (DEA) in the Peru Basin. A suite of state of the art methods were used to characterise physicochemical changes of the benthic habitat as well as effects on benthic fauna and biogeochemical processes. Towed, remotely controlled, and autonomous platforms equipped with high-precision navigation equipment allowed for precise habitat mapping, subsequent sampling, and in situ measurements directly at the seafloor, addressing specific levels of disturbance (e.g. specific impact features within the tracks). Results obtained so far indicate long-lasting disturbance effects, which were found to be statistically significant, despite spatial heterogeneity in the composition and biomass of benthic communities. Besides the disturbance of seafloor integrity, surface productivity and specifically nodule coverage were identified as key factors influencing benthic community activity and distribution. This calls for preservation zones that match exploitation areas in terms of nodule coverage and other physical habitat characteristics (e.g. topography, sediment properties). Likewise, the spatial planning of reference areas and monitoring strategies should focus on similar and ideally proximate environments in order to produce representative baselines and to properly identify the environmental impacts of mining operations.

Effects on seafloor communities and functions

Local comparisons of disturbance tracks with nearby reference areas identified effects on all benthic size classes, including microbial communities, and at all investigated temporal scales (from a few days to > 30 years). Taking into consideration that the spatial extent of such experimental disturbances and their local intensities are small compared to industrial-scale mining operations, the results indicate that long-lasting impacts from nodule mining have to be expected. Nodule removal and the associated alteration or loss of the upper sediment layer reduced abundances of the smallest benthic organisms (microorganisms and meiofauna (< 300 μm in size)). Many sessile, nodule-attached organisms, as well as mobile species did not reach pre-disturbance abundances decades after the disturbance took place. Likewise, the biogeochemical functions of the sediments were clearly affected. Laboratory and *in situ* measurements show reduced rates of organic matter remineralisation as well as lower microbial activity and growth in disturbed areas, indicating effects at the base of the benthic food web. High-resolution studies of habitat characteristics and functions in the DEA identify the loss of seafloor integrity as a particularly important factor. Fauna and processes are affected where the top few centimetres of sediments that developed over a period of thousands of years were lost or mixed by ploughing. The strongest effects were observed where the reactive surface layer was removed or covered with organically-poor, less porous, and stiffer sub-surface sediment.

Needs for future work and recommendations from existing studies

Clearly a lot more work is needed to assess the good environmental status of deep-sea ecosystems and the potential impacts of mining-related disturbances, specify acceptable impact thresholds, identify best-suited impact indicators, and to improve existing technologies for routine monitoring. Future studies would also need to shift from comparison of ecological patterns at the scale of local disturbance tracks to impacts across larger areas (e.g. equipment testing patches, license areas). Besides the assessment of direct impacts on seafloor integrity, it is important to study the effects of secondary impacts due to the deposition of fine-grained material settling out of sediment plumes that are created by nodule extraction ("blanketing"), as well as the potential consequences of the deposition of nodule debris. In addition, studies on potential restoration activities (e.g. deployment of artificial substrates, nutrient enrichment) need to be initiated in order to test their potential role in facilitating the reestablishment of seafloor fauna and ecosystem processes. Based on the close connection between physical disturbances and effects on seafloor biota and biogeochemical processes identified by recent investigations, monitoring of seafloor integrity is suggested as a simple proxy to discriminate between 'good environmental status' and 'harm'. This may be largely based on high-resolution imaging: changes in seabed colour and roughness as well as significant sediment coverage on nodules and sessile organisms provide clear indications of seafloor disturbance. For the assessment of more subtle changes in seafloor physical properties, standard sample-based methods are available (e.g. porosity and shear strength measurements, radionuclide profiles, X-ray sediment scans). Likewise, simple incubation-based, biogeochemical analyses (e.g. extracellular enzymatic activity and microbial growth) enable monitoring of biogeochemical functions before operational and autonomous instruments for direct monitoring at the seafloor become available.

Take-home messages

- Nodule ecosystems harbour a highly diverse fauna: Infauna, nodule fauna and mobile fauna are affected by disturbance and nodule removal;

- The characteristics of reference / conservation areas need to match those of mined areas (e.g. productivity, nodule coverage);

- The high spatial variability in faunal communities and functions means that detailed, site-specific investigations are necessary;

- Effects of (small-scale!) disturbances on nodule ecosystems last for decades and include all ecosystem compartments, including fauna of all size classes as well as biogeochemical ecosystem functions;

- Seabed integrity is a simple proxy for 'good environmental status'; visual seafloor disturbance is a simple proxy for 'harm';

- Appropriate methods for monitoring mining impacts and effects are available;

- Little is known on the fate and effects of mining plumes (suspended and dissolved matter) and tailing deposits;

- Research on restoration principles is required.

5.4.2 Significant, Serious and Sobering: Defining Serious Harm and Harmful Effects from Seabed Mining
Lisa Levin

This presentation summarised the output from two Deep Ocean Stewardship Initiative workshops held at Scripps in March 2014 and February 2017 and from the manuscript Levin et al. (2016)[1], in which the definition of harmful effects and serious harm were discussed. The mandate for considering these thresholds comes from UNCLOS, with the requirement that the ISA to adopt rules, regulations and procedures to "ensure effective protection of the marine environment from harmful effects arising from seabed mining". The potential for serious harm in a specific mining claim may be sufficient to trigger disapproval of a contract or emergency orders to alter or terminate operations, and may trigger liability for harm. Serious harm is mentioned in at least 11 sections of the ISA Discussion Paper on the Draft EnvRegs.

Weighing serious harm against environmental goals

Key issues and challenges in defining serious harm include the fact that, based on the best available scientific understanding, *mining causes serious harm* in the direct footprint. Thus, if mining is to be allowed, regulators may need to be prepared to authorise a fixed amount of harm – the key question is how much? Both context dependence and cumulative effects must be considered. A fixed amount of habitat degradation may be serious in one place and not in another, or as one of multiple mining actions, but not alone.

The assessment of harmful effects and serious harm must be weighed against a set of environmental goals and objectives. Six overarching environmental goals and eight detailed objectives for one of these were suggested as a basis for harm assessments. Recommended goals invoke viewing the environment as the common heritage of mankind (to be preserved for future generations), the concepts of sustainable development, precautionary measures, ecosystem integrity, context-specific environmental management, generating and sharing best available scientific information, and integration of strategic and contractor environmental management plans. With respect to the goal of sustaining benthic and pelagic ecosystem integrity, objectives should include ecosystem protection from contamination by pollutants, maintenance of population replacement through genetic connectivity, the preservation of a suitable habitat, sustained ecosystem function, the maintenance of genetic, species, habitat and structural diversity, sustained ecosystem services, the promotion of resilience, the incorporation of uncertainty into risk assessment, and context-specific environmental management.

The target ecosystems (nodule provinces on abyssal plains, massive sulphides at active and inactive vents, and ferromanganese crusts on seamounts) and associated functions and services were described. The harmful effects of seabed mining, to be avoided in order to protect the marine environment, were deemed likely to occur as a result of (a) removed or altered substrate (including topography, heterogeneity, texture), (b) altered geochemistry, (c) loss of biogenic habitat, (d) introduction of suspended sediments and contaminants, (e) loss of connectivity when habitat spatial extent is limited, and (f) the life-history attributes (longevity, slow growth) associated with deep-sea taxa, that will limit recovery and resilience.

---

[1] Levin, Lisa A., Kathryn Mengerink, Kristina M. Gjerde, Ashley A. Rowden, Cindy Lee Van Dover, Malcolm R. Clark, Eva Ramirez-Llodra, Bronwen Currie, Craig R. Smith, Kirk N. Sato, Natalya Gallo, Andrew K. Sweetman, Hannah Lily, Claire W. Armstrong, Joseph Brider (2016). Defining "Serious Harm" to the marine environment in the context of Deep seabed Mining. Marine Policy 74: 245-25.

Operationalising the definition of serious harm

Operationalising serious harm will require defining relevant time scales and considering harm over regional space scales within biogeographic provinces. For example, within a meta-community formed of patches that exchange individuals and species, serious harm may occur when losses prevent the persistence of specific taxa or when significant declines in critical ecosystem functions occur. Criteria for serious harm should be developed across multiple dimensions (e.g. an ocean health index) and should incorporate precaution and flexibility to account for currently unknown sources of serious harm. Harmful effects will happen at the species, community and ecosystem levels; alone or in aggregate they may reach a threshold of serious harm. Such thresholds will require definition.

To assess serious harm, multiple indicators will need to be monitored. These include: loss of biodiversity, species composition, ecosystem engineers and foundation species, habitat types, heterogeneity, endangered species, connectivity, productivity, respiration, nutrient cycling, trophic structure, demographic structure, recovery, and resilience. Key research gaps include understanding the regional distribution of habitats (active and inactive vents, seamounts, other features), natural variability, connectivity, succession and endemicity of taxa, the ecotoxicology of plumes and their interactions with fish and fisheries (seamounts); faunal sensitivity to changes in substrate and chemistry as well as impacts within the water column and at the surface. Detection of harmful effects or serious harm will require that a strategic environmental assessment is completed and protected no-mining areas are put in place *prior to* awarding any additional contracts; a sound design of a network of Preservation Reference Zones and Impact Reference Zones; and broader regional sampling outside contractor mining claim areas. Ultimately, the burden of proof should be placed on the proponent and the ISA to demonstrate a reasonable trade-off between benefits of mining to humankind and the costs, including non-economic costs, in the face of high uncertainty, high risk and the long-term nature of the harm.

In summary, the thresholds and tipping points that define serious harm will be a challenge to identify and should be linked to the precautionary principle. This will require a multi-dimensional, scientific approach. Such an approach may be developed by an expert advisory panel, which could be constituted by the ISA to define 'harmful effects' and 'serious harm' at all appropriate phases of environmental management.

5.5   Plenary Discussion of the Opening Session

In the discussion round after the welcoming remarks and key-note presentations, the point was made that technology could and should not be an element of the definition of serious harm, as constant change and progress in technological development may be expected to occur. The opinion was expressed that the type(s) of technology applied during mining operations can of course significantly contribute to reducing environmental impacts. In this regard, a view indicated that it is unlikely that impacts can be reduced to negligible levels and therefore, adequate regulations and codes of practice have to be developed.

The question was raised as to whether nodule mining may also have beneficial effects on the prevailing fauna, i.e. by creating niches for other fauna, or by enhancing the (re-)colonisation potential of organisms (e.g. through mining-induced dispersal). In addition, the importance of taking natural variability into account was emphasised. An opinion considered that the scientific approach is to measure and preserve species diversity and the complexity of food webs beyond their natural variability – and so far, all observations only point to a loss of species carrying out important functions, including the unique fauna associated with nodules. It is therefore imperative to leave some nodules on the seafloor and to install representative reference areas in the appropriate locations. Any beneficial effects of mining within the meaning of a benefit for humankind as a whole, would have to be defined in the context of environmental objectives, i.e. to preserve/conserve natural diversity and avoid loss of biological reserves (e.g. to preserve marine genetic resources as a pool for current or future use in medicine).

Is the recovery of minerals from the deep seafloor a necessary and responsible activity in view of the necessary transition of global economies to more sustainable societies as set by the global SDGs (in particular Goals 8 and 12)? Increasing resource efficiency, and turning to more sustainable production and consumption was seen as the long-term goal to be reached with an intermediate lack of supply which could be covered by either land mining or deep seabed mining.

The role of quantitative sampling was emphasised in order to develop regulations that neither over- nor underestimate the scale of environmental impacts. Yet, so far, scientific interpretation of disturbance-related impacts was only possible on a very small scale, and these do not allow extrapolation to the effects of a commercial-scale activity. So far, no experience can provide advice on the scale of sampling required for representative monitoring of impacts. The view was raised that a concept for a normative approach to serious harm is lacking.

It was suggested that the replacement of artificial substrate and potentially also of organic carbon (to kick-start re-establishment of the bioactive layer of sediments) may significantly enhance the recovery of the disturbed sediments. However, such experiments will require decades to provide proof of success and should be started immediately. Even in well-known ecosystems, restoration is extremely difficult, and so any restoration measures should be tested thoroughly prior to using them as a management tool.

6. Basis for the Workshop: Overview, Critical Analysis and Gap Analysis of the Discussion Paper on the ISA Draft Environmental Regulations

6.1 Presentation of the Discussion Paper on ISA Draft Environmental Regulations (*Chris Brown*)

6.1.1 Introduction and Background

The purpose of the presentation was to provide a broad overview of the ISA's Discussion Paper on the development and drafting of Regulations on Exploitation for Mineral Resources in the Area (Environmental Matters)[2], issued by the Secretariat in January 2017, and to outline the structure, challenges, and gaps in the development of the paper and tentative Draft EnvRegs.

It was noted that the main goal in the development of the EnvRegs is to develop a set of regulatory provisions that ensure the approval process for a Plan of Work for exploitation fully integrates environmental considerations and that such considerations continue throughout the life (including closure) of a mining project in the Area; and particularly, to: -

- contribute to the delivery of Article 145 RRPs requirement (effective protection of the marine environment from harmful effects);

- prescribe for key procedural obligations with regard to environmental assessment and management;

- deliver expectations for applicants as to documentation requirements / evaluation process;

- build on requirements under existing Exploration Regulations / exploration contract; and

- further promote access to information & consultation in the environmental process.

The Discussion Paper and the Working Draft (2016) have drawn on the provisions of the Convention and good national practices together with a wealth of workshop outputs and inputs from the ISA's stakeholder base over the last 3 years. Participants were asked to consider the key questions posed at paragraph 13.1 in the Discussion Paper.

The significance of the precise interpretation of Article 145 was highlighted. Nevertheless, clearly defined objectives to operationalise this article are required in order to formulate environmental targets (being an appropriate mix of qualitative and quantitative targets) for the purposes of monitoring appropriate management responses and targets against which to assess a contractor's environmental performance; that is, the delivery of an outcome-based approach to the environmental management of activities in the Area.

---

[2]   Available at https://www.isa.org.jm/files/documents/EN/Regs/DraftExpl/DP-EnvRegsDraft25117.pdf.

6.1.2 Structure of the draft Environmental Regulations

It was noted that the current Draft is split into sixteen parts and annexes, with varying degrees of content population. To assist in the understanding of an application and approval process, workshop participants had been provided with a process flowchart prepared by a working group of the Deep Ocean Stewardship Initiative. A copy of that flowchart is shown in Figure 1.

Flowchart of application process in the discussion paper for environmental regulations

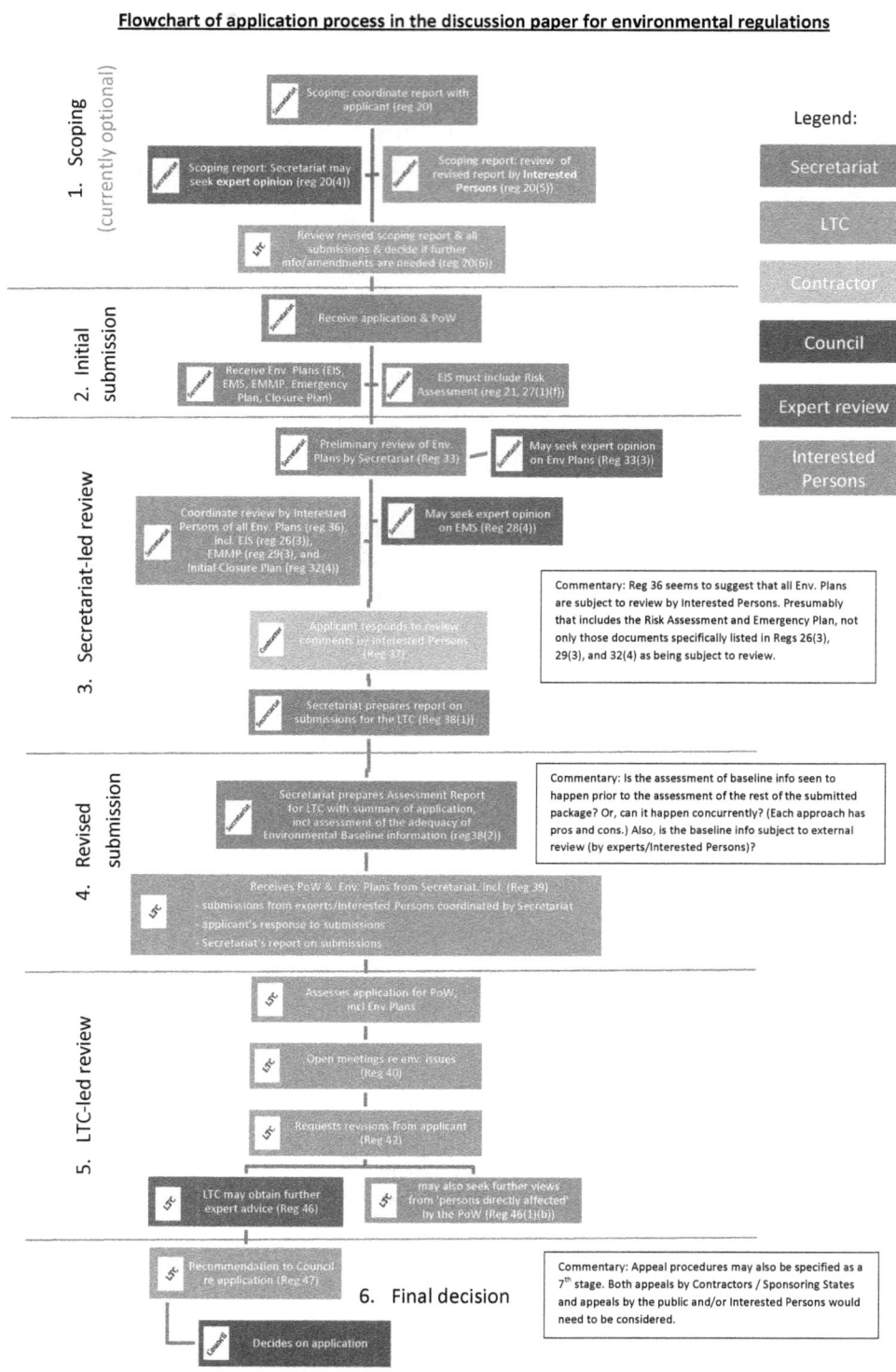

Source: Deep Ocean Stewardship Initiative (DOSI), Working Group 'Minerals' (March 2017)

Figure 1: *DOSI Flowchart of the application process as set out in the ISA Discussion Paper on the Development and Drafting of Regulations on Exploitation for Mineral Resources in the Area (Environmental Matters).*

Part II to the Discussion Paper on the Draft EnvRegs contains a number of guiding values that flow through to other regulatory provisions within the draft. These include: best available scientific evidence; access to environmental information; consultation in environmental decision-making; an obligation to deploy best environmental practices (BEP); and the application of the precautionary approach. It was noted that BEP, as a dynamic principle, requires the development of suitable criteria rather than a static definition. As to the application of the precautionary approach, one key feature in the draft is the requirement for an applicant to identify where the precautionary approach has been considered and to state associated precautionary measures (draft regulation 32(1)(r)); equally, in assessing an application, the Legal and Technical Commission are also obliged to state where the precautionary approach has been used in making its recommendations (draft regulation 47(1)(l)).

An outline of the constituent Parts to the draft was presented and a number of points and key themes were highlighted, together with those Parts requiring further consideration and analysis including Part IX (environmental management and monitoring), Part X (social and cultural management) and Part XII (compensatory measures). Additionally, further work is required to draft environmental management terms for the standard exploitation contract.

6.1.3   Challenges and Gaps

The following points were highlighted during the presentation for further attention:

- Some regulatory provisions require input from technical experts. Draft regulation 21 relating to risk assessment was presented as an example of this;

- Timelines: the importance of a responsive and time-bound decision-making process was iterated and it was noted that the draft exploitation code as a whole would reflect more precise timelines in due course.

- The "Authority": the current generic references to the ISA in the regulatory framework will be amended as the day-to-day functioning of its respective organs are clarified, including that of a staff of inspectors;

- Stakeholder expectations: a clear process and procedure for stakeholder consultation should be outlined in order to manage stakeholder expectations effectively;

- Role of sponsoring States: the role and function of sponsoring States is yet to be fully articulated. It is intended that a matrix setting out the duties and responsibilities of the ISA and sponsoring States together with that of flag States will be prepared following appropriate consultation;

- "Confidentiality": it was noted that the stakeholder responses to the Working Draft (2016) showed diverging views on the definition of confidential information as being potentially too broad or too narrow. One proposal under consideration is to compile a comprehensive list of information that will be treated as non-confidential;

- Technical and economic constraints: measures should be assessed against reasonable technical and economic constraints, though this phrase requires further elaboration;

- Cost of compliance: the costs relating to the implementation of the regulatory provisions will need to be evaluated;

- Definitions: it was emphasised that the content of the regulatory provisions should drive the definitions. As to the challenge of defining "serious harm", while the phrase merits development of suitable criteria, the 1990 Prep Com definition (see para. 7.6, page 10 of Discussion Paper) provides for any effect "beyond that which is negligible or which has been assessed and judged to be acceptable by the Authority". This element of that definition points toward a need to establish what is / are acceptable effects that can be used for the assessment and evaluation of an application in the first instance;

- Terminology: it is recommended that regulatory language and terminology should mirror that in the Convention. Other terminology should be driven by need and content. Confusion relating to terminology proposed for strategic/regional management and planning highlighted the challenge here and a requirement to keep document needs simple for ease of understanding and implementation.

Finally, it was noted, for reasons of clarity and integration, that the environmental provisions would likely be combined with the existing Working Draft (2016) to form a consolidated set of regulatory provisions for exploitation activities in the Area.

6.2 Five Critical Statements on the ISA Draft Environmental Regulations from Different Stakeholders

6.2.1 "Four Critical Statements": A Personal View – and NOT by the LTC
Christian Reichert

The following statements have been made by C. Reichert (Chairman of the LTC) in his own personal capacity. There is no authorisation by the LTC, amongst others as negotiations and discussions within the LTC are still in progress and have not been finalised.

The "Regulatory Framework for Mineral Exploration in the Area" consists of General Provisions, Environmental Matters, the Mining Directorate and possibly further elongated chapters. All this is discussed under the pre-condition that: NOTHING IS AGREED UPON UNTIL ALL IS AGREED UPON!

Statement one

The process under development on how to assess deep seabed mining in terms of environmental impacts, and how to decide on approval, change/amendment, or rejection is an utmost complicated one that easily leads to misinterpretation or even failures when operationalising in the real world. Hence, a clear glossary and a clear roadmap in terms of content, hierarchy and time line has to be urgently developed as one of the very first steps (comprising, amongst others, definitions of terms such as serious harm, adaptive management, precautionary approach/principle, EIA, SEA, environment baseline studies, etc.).

Statement two

The Manganese Nodule Belt (CCZ) - as an example - stretches over a large region whereof only limited areas are under contract or protected as APEIs. However, organisms and their ecosystems are not confined by arbitrary boundaries. Hence, it follows that huge research efforts have to be made in the APEIs and in the interstices between contracted areas and the APEIs. Thus, strong cooperation between all stakeholders is highly recommended, and international as well as national research funding agencies should be involved.

Statement three (contactor's workshop, Kingston, Jamaica, July 2015)

Before commencing into the full production phase, the following requirements have to be fulfilled for the benefit of the regulator and the entrepreneur:

- Proof of full technical feasibility (fully-fledged pilot mining test (PMT), subsequent production test);

- Complete business/financial models, taking into account the entire value chain;

- Comprehensive risk models including technical, environmental (liability), and economic/financial risks.

Hence, it follows that a demonstration project is urgently needed. Also in that phase, far-reaching cooperation between all stakeholders is highly recommended. A useful example might come from the oil and gas industry, in which there is almost no pioneer project which is not secured by a consortium of companies sharing the risks and the benefits equally. It has to be kept in mind that the time line from PMT preparation with the regulator to the final conclusion of the test and its environmental monitoring stretches over some three to five years!

Statement four

The main issues that need to be swiftly solved are already covered in the ISA Discussion Paper, Section I, pt. 13.1 (pg. 17).

6.2.2   Critical Statements on Contents and Structure by an NGO
Duncan Currie

Introduction and objectives

Article 145 UNCLOS sets out the ISA's mandate to adopt rules, regulations and procedures to prevent, reduce and control the following:

- Pollution and other hazards to the marine environment;

- Interference with the ecological balance of the marine environment;

- Protect the marine environment from harmful effects of "activities in the Area";

- Protect and conserve the natural resources of the Area; and

- Prevent damage to the flora and fauna of the marine environment.

These tasks are implemented though Article 165 (2) UNCLOS, according to which the Legal and Technical Commission (LTC) shall make recommendations to the Council on the protection of the marine environment, take into account the views of recognised experts, and formulate and submit regulations and procedures to the Council taking into account assessments of the environmental implications of activities in the Area. Pursuant to this, the Council shall adopt and, pending approval by the Assembly, provisionally apply the regulations, taking into account the recommendations of the LTC or other subordinate organs as necessary.

In accordance with Article 160 (2) UNCLOS, the Assembly considers and approves regulations which uphold the objective criteria contained in Annex III, Article 17 UNCLOS. These criteria include operational safety, resource conservation, and the protection of the marine environment. The latter includes measures to secure the effective protection of the marine environment from harmful effects directly resulting from activities in the Area, including effects from shipboard processing of minerals immediately above the site where those minerals were extracted, taking into account the extent to which harmful effects may directly result from drilling, dredging, coring and excavation, and the disposal, dumping and discharge of sediment, wastes and other effluents into the marine environment.

Noting the reference to subsidiary organs in Article 162 (2)(o) UNCLOS, Article 158 provides that subsidiary organs may be established as necessary in accordance with Part XI. The 1994 Implementing Agreement further sets out that the establishment and functioning of the ISA's organs and subsidiary bodies shall be based on an evolutionary approach. The 1994 Implementing Agreement directs the ISA to emphasise the adoption of rules, regulations and procedures for the protection and preservation of the marine environment, as well as the timely elaboration of rules, regulations, and procedures for exploitation, including those relating to the protection and preservation of the marine environment in this context. Once these regulations have been developed, applications for plans of work for exploitation are submitted for consideration by the Council on the basis of recommendations made by the LTC, in accordance with Article 153 (3) UNCLOS.

Recommendations on operationalising the Regulations

Essential elements of the regulations include substantive requirements addressing environmental baselines which account for all impacts, including cumulative impacts; strategic environmental management plans (SEMPs) for all areas; the creation of protected areas; effective management; as well as liability and redress. Procedural requirements include transparency, public participation, and review procedures.

Concerning transparency, the Aarhus Convention, which was incorporated into the environmental management plan for the Clarion-Clipperton Zone, has three "pillars": access to information, public participation and access to justice. The Rio+20 outcome document *The Future We Want* stated the need for institutions at all levels that are effective, transparent, accountable, and democratic. This implies a need to build in robust, and transparent public participation mechanisms addressing review, dispute resolution and appeals, as well as access to environmental information. Public participation requires clear, transparent timelines; comprehensive and fair evaluation of applications; independent scientific review of EIAs; public comment and review procedures; as well as regular reviews. Further, the Almaty Guidelines on promoting the application of Aarhus principles in international for a call for international organisations to develop and publicise clear and transparent policies and procedures enabling access to environmental information.

The EIA process set out in the regulations needs to be transparent; include public review and comment; ensure scientific peer review; allow all stakeholders to ask questions and seek further information; and ensures open hearings at which stakeholders are allowed to seek further information, question scientists, and present independent evidence.

The Discussion Paper requires additional components including strategic environmental assessment, regional environmental management plans, review and appeal procedures, and liability and sustainability funds. One important issue is establishing the threshold for "effects" to be operationalised in the regulations: the term "serious harm" is referred to only in Article 162 (2)(w) UNCLOS regarding emergency orders , Article 162 (2)(x) regarding disapproval of areas for exploitation, and in Article 290 concerning the threshold for provisional measures. In contrast, Article 145 UNCLOS uses the term "harmful effects". It is recommended that the threshold of "serious harm", defined in terms of "significant adverse effects", be operationalised concerning intervention and liability, but that the regulations use the term "harmful effects" and the other criteria contained in Article 145.

On the issue of legal definitions, the term "acceptable" should not be defined in regard to financial consequences. Further, the term "adaptive management" should adopt the criterion used in the New Zealand Supreme Court case *Sustain our Sounds*, where the question was raised whether any adaptive management regime can be considered consistent with a precautionary approach and the key criterion set out was the extent to which adaptive management reduces uncertainty and risk.

The proposed definition of "interested persons" is inappropriate when considered in light of the principle of the common heritage of mankind, which is premised on the potential interest of all mankind. The corresponding definition used in the Aarhus Convention of "public concerned" specifically includes environmental NGOs and members of the public with interest in the subject matter.

It is also recommended that economic considerations be deleted from the Draft Regulations, following on account of several passages in the ITLOS Advisory Opinion: 1) the operational standard for due diligence requires more stringency for riskier activities (para. 117); the level of scientific knowledge and technical capability in the necessary fields available to a given State, as opposed to generally available, is the decisive factor (para. 162); as well as the requirement to uphold a precautionary approach and apply best environmental practices (para. 242).

6.2.3 DOSI Comments on the ISA Discussion Paper on the Draft Environmental Regulations - A Science Perspective
Eva Ramirez-Llodra and the DOSI team

Background

In February 2017, DOSI (Deep Ocean Stewardship Initiative) convened a workshop[3] to discuss elements of the ISA Discussion Paper on the development and drafting of Regulations on Exploitation for Mineral Resources in the Area (Environmental Matters). The overall objective of this DOSI workshop was to evaluate and review the ISA Discussion Paper from a (predominantly) scientific perspective. DOSI has in the past submitted comments on previous ISA documents (the Regulatory Framework in 2015 and the Working Draft for Exploitation Regulations in 2016). For regulations focusing on environmental matters, the workshop provided an initial step; DOSI plans to formulate a comprehensive response when the request for public input

---

[3]   1-3 February 2017, <u>Workshop Co-Leads</u>: Lisa Levin & Verena Tunnicliffe; with 18 participants from nine countries

comes. The ISA has a huge and difficult task ahead and the DOSI Minerals WG members expect to help, alongside others, to ensure significant (and crucial) scientific expert input at an early stage.

General response to the ISA document

The ISA Discussion Paper provides a good starting point for discussion and input to very difficult questions, many of which require significant deliberation in order to develop clarity within the regulations. The Draft provides a comprehensive outline for taking a contract application from submission to approval. Elements of particular strength are recognition of: (a) the balance required between strict regulations and flexible operations, including the need to include procedures for adaptive management; (b) a pathway for scientists to participate and provide advice for knowledge-based decisions; (c) appropriate review of applications for exploitation, both internally within the ISA and with external experts, as well as consultation with Interested Persons; (d) ecosystem-based environmental management, underscoring the need for integrated and holistic approaches; (e) the importance of protecting Vulnerable Marine Ecosystems and habitats, consistent with the Exploration Regulations; (f) the valuable combination of a scoping exercise and an EIS as components of the EIA; and (g) importance of transparency and public participation as well as open data sharing. Issues needing further attention were identified, discussed and some initial feedback was delivered to the ISA (Feb. 2017). Amongst other issues discussed, the group highlighted the need for a greater emphasis on the Common Heritage of Mankind (CHM) and the concept of "the benefit of mankind" as well as the need for further support on existing and new technologies so the remoteness of deep-sea ecosystems is not a limitation for monitoring.

Vulnerable Marine Ecosystems (VMEs)

How is the VME concept applied in terms of mining? The FAO has guidelines that apply to impacts on vent ecosystems by fishing. However, the spatial scale would be larger for vent mining than fishing, and the temporal scale will be longer for nodule mining than fishing. Further work needs to build on ongoing discussions (e.g. SEMPIA).

Definitions

DOSI has selected several key terms for priority attention, for which we have suggested alternative wording: Acceptable, Benthic Plume, Environmental Baseline, Environmental Impact, Environmental Target, and Impact Reference Zone. Input for the other definitions has also been provided and will be conveyed to the ISA.

Aims of the Discussion Paper on Environmental Regulations

The Draft Regulations indicate a shift in approach for the ISA that includes good public consultation and involvement of experts. Some terminology needs clearer definition. A subgroup, led by A. Jaeckel, produced a flowchart (see Figure 1) on the mining exploitation application process, from the scoping report to approval that can be used to aid the further formulation of robust planning.

The environmental baseline

A detailed list of recommendations was drawn up and included, amongst others: the ISA should develop a separate guidance document on the roles and responsibilities of sponsoring States reflecting the ITLOS

Advisory Opinion; Environmental Regulations should cover all deposit types, with deposit-specific additional Guidelines/Recommendations, as these may require more frequent amendment; Environmental Regulations should reference and align with SEMP; a Data Management System needs to ensure open reporting of all environmental data and DOSI suggests that all data collected by contractors is reported to the ISA 3 years post-cruise; workshops on data standardisation are encouraged and suggestions were made for training contractors on sampling methodology during an 'at-sea' practical workshop; best practice updates need to be made available to contractors on the ISA website.

Environmental goals

Environmental goals were referenced in the Discussion Paper, but no goals or objectives were stated. Thus, the DOSI WG discussed what those goals and objectives could be. A set of potential goals were identified, drawing on UNCLOS, FAO, and ISA precedents. Further work is needed, but initial goals can include: 1) preserve CHM for future generations, including biological, geological and cultural resources and services; 2) ensure that development of DSM is carried out in the context of sustainable development; 3) protect and preserve the marine environment through the application of the precautionary approach; 4) sustain marine benthic and pelagic ecosystem integrity, including the physical chemical, geological, and biological environment; 5) generate and share best scientific information available for decisions-making and improve techniques for dealing with risk and uncertainty; and 6) ensure ecosystem integrity on regional scales by integrating strategic and contractor environmental management plans.

Guiding principles

Recommended edits and inclusions were suggested, amongst others: the Preamble should include an objective to ensure a transparent decision-making process; the Preamble should include an objective to consider the CHM by not foreclosing on options for future generations; the term "ecological balance" should be defined; references to the precautionary approach should use the wording in the Advisory Opinion and its application to only Serious Harm situations be reconsidered (it needs application at a lower threshold).

Environmental Impact Assessment

Several points were made and included: the Scoping Report should be mandatory to help both Contractors and the ISA to identify issues at an early stage; include expert review at the scoping stage; consider an entry for the Scoping Report to include CHM; define thresholds for impacts; the plume is not the only indicator to determine the Environmental Impact Area and other processes such as population connectivity and source populations, nursery areas, migration routes, noise, trophic interactions and ecosystem functions should be included in the baseline and EIA.

Serious harm

The DOSI WG acknowledged that an operational definition of serious harm must underlie many aspects of the Environmental Regulations. The definition and operationalisation of serious harm were discussed, as summarised in CHAPTER 5.4.2.

6.2.4 Critical Statement by a Contractor
Ralph Spickermann

It is important to address environmental concerns in the context of UNCLOS, which sets the overarching legal framework for the responsible development of the resources of the Area. UNCLOS specifically provides for exploitation activities in the Area, with a view to ensuring the development of the resources; and that this activity is to be carried out in such a manner as to foster healthy development of the world economy[4]. Areas are to be disapproved for exploitation in cases where substantial evidence indicates the risk of serious harm to the marine environment[5].

Twenty Member States now sponsor ISA contractors. These sponsoring States represent 47% of human population and 43% of global Gross Domestic Product. These States span largest to smallest, developed to developing. Contractors do not comprise a global niche interest; they are sponsored by governments representing nearly half of mankind to be the actors to bring the resources of the Area into use. The sponsoring States comprise a growing and diverse global constituency for the responsible development of the resources of the Area.

The size of an ecosystem is a foundational consideration for any environmental analysis, particularly to determine the fraction which, if disturbed, would constitute serious harm. The geographic scale of seabed resources is vast. The Working Draft for Exploitation Regulations is silent on this key point and the fact that it would take years for an operation to disturb even 0.1% of the CCZ (Figure 2). This scale gives the ISA great opportunity and flexibility to continue to improve its ability to manage the CCZ. The ISA could modify (or cause contractors to cease) operations according to the standards in the exploitation code.

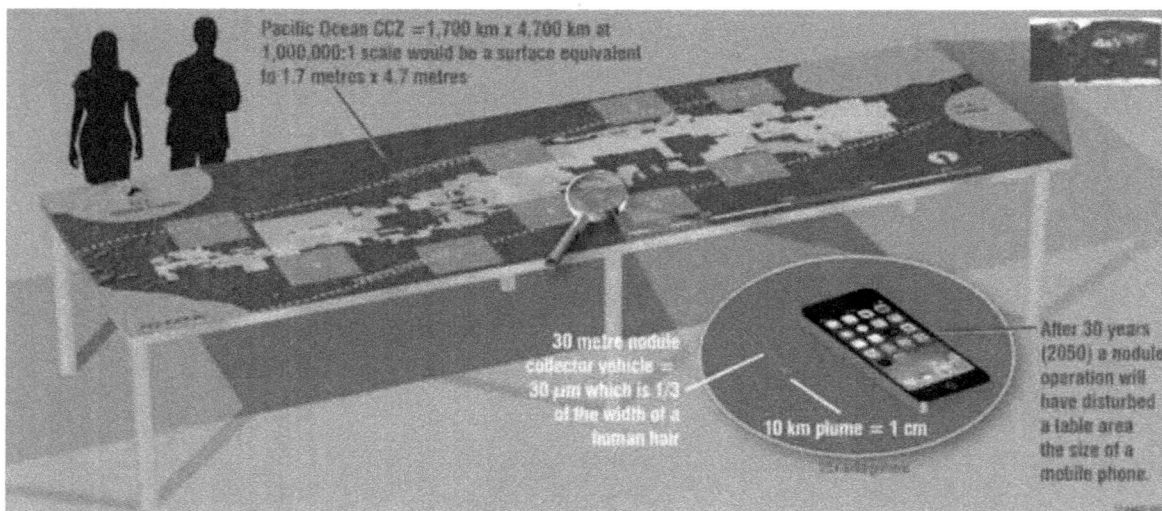

Figure 2: *Geographic Scale of Seabed Resources. As shown, a large conference table represents a 1,000,000:1 scale model of the Clarion-Clipperton Zone. After 30 years, a nodule exploitation operation will have disturbed only a small fraction. The vast geographic scale of seabed resources gives the ISA great flexibility to constantly improve its management, even during years of initial exploitation.*

---

[4] UNCLOS Part XI Article 150
[5] UNCLOS Part XI Article 162

The regulations should take into account the vast size of the Area, designated as the common heritage of mankind and proportionally vast ecosystem, the consequential low risk of immediate serious ecosystem harm, and the scientific hypothesis[6] that the only way to rigorously determine the consequences of full scale operations is to build and operate them. Therefore, the regulations should focus on criteria for localised environmental baselines and associated scientifically determined preservation zones, and then monitoring/learning/managing operations for such time as (1) they make economic sense for the contractors and (2) the extent of unavoidable environmental consequences are firmly established. Only by observing and managing full scale operations for some years can mankind decide if the benefits are outweighed by the environmental consequences that necessarily ensue (Figure 3).

The regulations must be fact-based, and each line should be scrutinised to assess whether it contributes materially to protecting the environment from serious harm, pollution or waste. An environmental management strategy for the Area is not called for in UNCLOS, and introduces an undue complexity, which leads to unwarranted burden in terms of capacity, fiscal and administrative resources, for both the ISA and the LTC, as well as the contractor itself.

The "do nothing" option must be included in the environmental assessment. That is, the clear and present risks of continued exclusive reliance on diminishing terrestrial sources. These include higher prices for cobalt/nickel/manganese leading to more expensive electric vehicle/renewable energy technology (and thus delayed adoption of renewables and increased global emissions); the relative carbon footprint of terrestrial mining operations; the regional scale watershed pollution/degradation due to mines in arid and rainforest landscapes; human deaths/birth defects/disabling injuries (to both children and adults) from artisanal mining of cobalt, manganese and rare earths; deforestation and water monopolisation; and associated social disruption. In addition, these risks put reliance on terrestrial resources under pressure from national considerations (e.g. Philippines[7], El Salvador[8], Chile, and the EU[9]).

---

6   U.S. NOAA Deep Seabed Mining, Final Programmatic Environmental Impact Statement, Section II.C.2.1, II.C.2.2 (1981)
7   Republic of Philippines Department of Environmental and Natural Resources, Final Mining Audit Report, 2 Feb 2017
8   http://www.reuters.com/article/us-el-salvador-mining-idUSKBN1702YF, 29 Mar 2017
9   Raw Materials Scoreboard, European Union, 2015 ISBN 978-92-79-49478-9

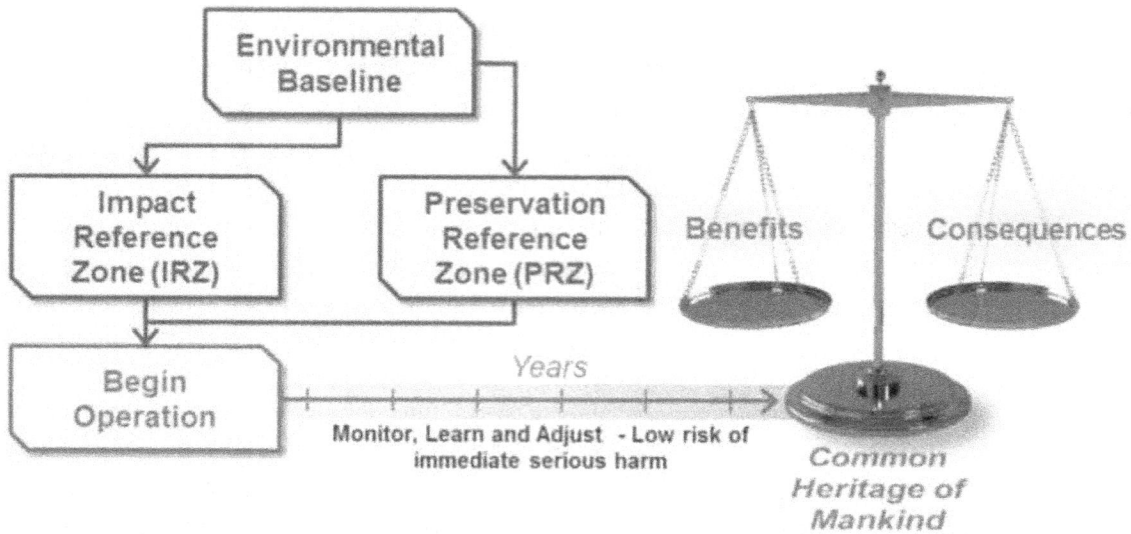

Figure 3:   *Fact-Based Multi-Generational Decision. UNCLOS mandates that regulations for polymetallic nodules be developed first. The LTC has acted accordingly by drawing up a CCZ Environmental Management Plan (ISBA/17/LTC/7). The Secretariat has drawn up next steps for implementation (ISBA/22/LTC/12) which include establishing IRZ and PRZ guidelines. These guidelines are the foundation for the contractors to complete their environmental baseline, choose initial exploitation sites, define associated IRZ and PRZ, obtain an exploitation licence and begin operations. Monitoring and adjusting active operations is the way the actual benefits/consequences can be determined and weighed, both for this generation and those to follow.*

The over $1 Billion of research since the 1970s has produced a portrait of the nature of the landscape, faunal classes, and their potential vulnerabilities. The ISA has a CCZ Environmental Management Plan in place. The regulatory drafting process should build on this existing plan and focus on implementing environmental requirements for polymetallic nodules. Once exploitation regulations are in place, and exploitation contracts issued, the ISA should, as operations proceed, work with contractors to monitor and adjust their respective operations (Figure 3). The near-term risk of ecosystem serious harm is mitigated by the geographic scale of the CCZ.. In this way, the actual benefits/consequences of polymetallic nodule exploitation can be determined and weighed, both for this generation and those to follow.

6.2.5   Critical Statement of the German Environment Agency
Harald Ginzky

The German Environment Agency (UBA) as a governmental, environmental regulator, is of the view that the Draft EnvRegs as summarised in the ISA Discussion Paper is too narrow in scope. As we are anticipating that projects may enter into the exploitation phase in the near future, a new dimension of potentially negative effects on the marine environment is to be expected and this has to be dealt with. It is the mandate of the ISA to tackle this challenge in a responsible manner. The Draft EnvRegs focus strongly on enabling exploitation projects and somewhat neglect the long-term mandate of the ISA to implement its obligations with respect to the principle of the "common heritage of mankind", the "Sustainable development agenda", and the idea of an equitable sharing of benefits.

Thus, the Draft EnvRegs (and/or an additional set of regulations or policies) should establish an appropriate set of tools that break down the global agenda to the dimension of regional planning. Regional assessment and planning is a necessity in order to achieve the above-mentioned mandate of the ISA as this is the only tool which can enable the ISA to deal inter alia with cumulative effects, with conflicting uses and with alternatives. Thus, it is essential that the ISA develops a regulatory approach that, amongst others, also focusses on the contents, procedures, responsibilities, and funding of regional plans.

Furthermore, the Draft EnvRegs lack clarity on how to enforce provisions. Protection of the marine environment requires effective implementation and enforcement. The Draft EnvRegs so far do not sufficiently clarify the roles and functions of the sponsoring State and the ISA organs, as well as of the contractors themselves.

Several aspects in the Discussion Paper require further clarification:

- It must be guaranteed that species and habitats are effectively conserved. In order to achieve this objective, sufficiently large and ecologically representative areas must be excluded from exploitation projects. We would recommend to state this requirement as a cut-off standard for the approval of exploitation projects.

- Current knowledge on the ecosystems that may be impacted by deep seabed mining is very limited and so the ISA will have to deal with many uncertainties. However, the Draft EnvRegs do not foresee a cut-off option in the case that uncertainties are so large that they do not allow a sufficiently effective assessment of the impacts on the marine environment. A precedent in international law, namely in Annex 2 of the London Protocol, states that… "If this assessment reveals that adequate information is not available to determine the likely effects of the proposed disposal option, then this option should not be considered further." A similar requirement should be integrated into the Draft EnvRegs.

- The definition of "interested persons", which are to be involved in consultation is not sufficient. The definition in Schedule 1 of the Discussion Paper demands that persons are "directly affected" by the exploitation activity, and moreover that this should be the case "in the opinion of the Authority". The ISA should, however, ensure that all persons, private or public that are interested in the consultation process can be involved.

- The term "Best available scientific evidence" is not sufficient as it could be understood to exclude scientific work in progress or scientific knowledge that has only been published in grey literature. In the case of especially risky activities such as nuclear power generation or genetic engineering, German law requires that all scientific knowledge, even if still under dispute, has to be considered by the competent authority. This approach should also be a standard adopted for exploitation activities due to the high level of uncertainties involved.

- The provision on Environmental Impact Assessments appears to be overly complex. It would be helpful to distinguish between (1) the content of the assessment, (2) the way the outcomes of the assessment are integrated into the Environmental Management and Monitoring Plan, and (3) procedural requirements.

- The Draft EnvRegs in their current form in the ISA Discussion Paper, can only be applied together with the future "Exploitation Regulations". We suggest that it would be reasonable to incorporate the

Draft EnvRegs into the future Exploitation Regulations. In addition, more clarity on the specific requirements must be achieved.

Concluding remarks

The term "Discussion Paper" under which the Draft EnvRegs were published appears to be an appropriate one as many aspects still require a detailed discussion (contents, conceptual approaches, wording).

The need for a long-term environmental strategy and/or regional planning instruments, as well as for conceptual approaches, for example on how to deal with uncertainties is still under debate. Such questions are by no means pure technicalities, but follow political preferences. An in-depth political decision on such issues is therefore required. Taking this into account, the ISA may want to reconsider its ways of cooperation and negotiation. A stronger involvement of States Parties in the decisions on the basic design and contents of the Environmental Regulations would be helpful to increase "ownership" by the Parties. To this end, the ISA should rely more on legwork by working groups composed of delegated members of States Parties and observers – either by correspondence or by physical meetings.

6.3   Gap Analysis of the Draft Environmental Regulations

6.3.1   PEW Foundation Perspective
David Billet

The Berlin Workshop "Towards an ISA Environmental Management Strategy for the Area" was preceded by two smaller meetings supported by Pew Charitable Trusts. One of these was hosted by the Deep Ocean Stewardship Initiative (DOSI) at the Scripps Institution of Oceanography (see CHAPTER 6.2.3) and the other at the Pew Charitable Trusts office in London. The meetings reviewed the *Discussion Paper on the development and drafting of Regulations on Exploitation for Mineral Resources in the Area (Environmental Matters)* that was published on the website by the ISA in January 2017. The two meetings gave scientific, legal and management experts an opportunity to review and discuss the ISA Discussion Paper and to prepare for the Berlin workshop. A number of presentations arising from this work were made at the Berlin workshop.

The Pew Foundation perspective presented by David Billet focused on four main aspects that address Draft Regulation 41 dealing with matters to be taken into account by the LTC when assessing the submission of an EIS: 1) whether the concept of 'Vulnerable Marine Ecosystems' was suitable for the inclusion in regulations for deep seabed mining, 2) the need to align the new Environmental Regulations with Strategic Environmental Management Plans (SEMPs), 3) how expert advice might be operationalised in assessing Environmental Impact Statements (EISs) and Environmental Management and Monitoring Plans (EMMPs), and 4) whether restoration and offsetting actions should be included in the regulations as mitigation solutions.

Vulnerable Marine Ecosystems (VMEs)

International Guidelines for the Management of Deep-Sea Fisheries in the High Seas were issued by the UN Fisheries and Agriculture Organization (FAO) in 2009. The Guidelines include a definition of the concept of VMEs based on the life history traits of species and offer an Annex with examples of potentially vulnerable species groups, communities and habitats, as well as features that support them. The list includes hydrothermal vent communities and features such as seamounts; the former of possible importance to polymetallic sulphide mining and the latter to cobalt crust mining, in some instances. In these specific cases, therefore, the FAO Guidelines could be applied. However, for inactive vent sulphide sites and for polymetallic nodules and their associated sediments, focusing on VME criteria may not be suitable because all the fauna is vulnerable owing to their life history characteristics, especially due to low reproductive output and long generation times. Issues of rarity in the FAO Guidelines would be difficult to apply owing to the apparent, but misleading, rarity of many species, which is more likely to be related to the paucity of sampling at abyssal depths. New methods and regulations based on ecosystem-based management, representativity, and spatial planning are likely to be more effective tools.

Environmental Regulations and Strategic and Regional Environmental Management Plans (SEMPs/REMPs)

Draft Regulation 41 requires the ISA to consider exploitation applications in the context of SEMPs and REMPs. These plans should be in place before applications for exploitation are made so that 1) contractors can set their EIS within a wider regional context, and 2) the ISA can consider the application in relation to issues such as cumulative impacts and environmental variability. The ISA has responsibility in generating the SEMPs and REMPs. The challenge is that the SEMPs and REMPs are dependent on environmental baseline data collected and reported by contractors during their exploration activities. At present much of this data are not available. There are considerable differences between contractors owing to how they have programmed their work during the exploration phase. The Draft EnvRegs offer an opportunity to specify the data to be provided by contractors in a timely fashion so that the ISA can fulfil its obligations to protect and preserve the marine environment.

How might expert advice to the ISA be operationalised?

The review of environmental baseline data submitted to the ISA will require quality checking. In addition the assessment of complex and extensive EISs and EMMPs will require specialist advice. In addition the review of an EIS may require expert assessment of the data submitted and upon which the EIS relies. Neither the Secretariat nor the LTC have the range of skills necessary to review the data presented in EIS and EMMP, which will include a wide variety of topics, such as physical oceanographic modelling of plumes, ecotoxicology, geochemistry, biodiversity indices, and measures of ecosystem functioning faunal groups from microbes to megafauna and in both the benthic and pelagic realms. Impartial assessment will be required, possibly by calling upon individuals on a case-by-case basis, drawn from a pool of accredited experts appointed for 5 years and taking into account geographic location and potential conflicts of interest. The Regulations may need to stipulate how expert advice will be sought and used by the ISA. The EIS, reviews by experts, responses by the contractors, and any arbitration between experts appointed by the contractors, the ISA and Interested Persons will need to be reported in full in order to gain a social license to operate.

Restoration and Offsetting

The sequential environmental management procedure for mitigation actions of 1) avoid, 2) minimise, 3) restore and 4) offset should be adopted. In many cases, it will be difficult for deep seabed mining to avoid impacts on the marine environment. There will be a wide range of mining footprints but all can be minimised through engineering innovations, the planning of mining activities and the use of spatial planning to create set aside areas for the preservation of the marine environment. For polymetallic nodules, contractors are already considering ways to reduce sediment compaction and the spread of seabed and discharge plumes. As in shallow water habitats, there is increased interest in how simple engineering approaches might assist the environment in recovering more quickly over time. This may be particularly important in abyssal areas where recovery rates are considered to be very slow. Restoration actions may include the placement of false nodules on the seabed as a substrate for sessile fauna which occur only on hard substrates. This may require the prior reconsolidation of the sediment surface through stimulating the production of exopolymers by microorganisms. Abyssal sediment communities may be induced to recover ecosystem functioning and structure through greater detrital organic inputs. Hydrothermal vent ecosystems may be assisted by introducing man-made sulphide chimney-building structures. Direct experimentation is required on potential restoration solutions during benthic impact experiments and mining tests. Offsetting in comparison may not be suitable because other deep-sea ecosystems are unlikely to require restoration actions. However, there are some suggestions that damage to deep-sea ecosystems could be offset in shallow water, for instance by rebuilding coral reefs. It may be necessary to include mitigation procedures, including restoration, explicitly in the Environmental Regulations.

6.3.2   Plenary Discussion after talk by David Billet

In the discussion, the VME theme was taken up. It was mentioned that for deep-water fisheries, the UN Regulations and the associated FAO Guidelines set an internationally accepted and globally applied framework for managing fishing activities. Different regions use different lists of VME indicator taxa. The use of these indicator taxa was not meant to prohibit all fishing but to ensure that fishing does not proceed in areas where "significant adverse impacts" on marine ecosystems have been identified (see FAO Guidelines, 2009). Indicator taxa are but one tool. One view was expressed indicating that the level of environmental protection provided through deep seabed mining regulations should be in line with those applied to deep-water fishing; and that; otherwise either non-compliance or a lowering of standards in fisheries could follow.

The issue was raised whether using presently incomplete lists of VME indicator taxa and habitats is an appropriate method to determine what is important in an ecosystem context, such as its underlying, potentially spatially explicit functions (nurseries, source or sink of a population, feeding areas), diversity, species interaction, etc. What exactly is an ecosystem and how can the concept be operationalised in space and time? In this regard, it was expressed that the qualifiers to be developed for characterising VMEs in the deep seabed mining context should be broader than mere taxa lists and include descriptors for ecosystem functioning. A view considered that for hydrothermal vents and seamount habitats, the approach from deep-water fisheries may be suitable nevertheless.

6.4   World Café Discussions

6.4.1   Is the Structure and Content of the Draft Environmental Regulations Adequate / Fit-for-Purpose?

The discussion at this World Café table considered the structure and content of the ISA Working Draft for Exploitation Regulations (2016) as well as of the Discussion Paper on the Draft EnvRegs.

The Discussion Paper on the Draft EnvRegs was generally considered adequate and fit-for-purpose. However, it was felt that the draft only provides a general structure concerning the necessary elements of final Regulations and require concretisation.

Structure of the Draft Environmental Regulations

One central outcome was that the scope and limits of the Regulations should be clarified. This would include the following topics:

- Clear distinction between exploration and exploitation;

- Clear definition of the limits of the ISA's responsibilities both in geographical terms (EEZ and the Area) and in substantive terms (clear definition of the term "activities in the Area"), including the definition of the roles and functions of different international bodies such as the IMO;

- Clear distinction between the responsibilities of contractors and the sponsoring States;

- Clarification of the roles and functions of the various ISA organs, including a decision on the need for additional bodies such as an Environmental Commission;

- Integration of all legal provisions into one document in order to improve accessibility and ease of use for smaller administrations and the general public. The structure of the Regulations should also be such that updating is facilitated;

- Clarification of the interplay between the various procedural steps and the various players. The flowchart provided by DOSI (Figure 1) was considered to be a helpful tool.

Content of the draft Environmental Regulations

The following aspects were considered essential:

- Definition of timelines, requirements, and obligations;

- Determination of all relevant issues to be regulated. Specific requirements could be determined by environmental standards, e.g. guidelines;

- Inclusion of a compensation scheme;

- Clearer "definitions" to be used in and by the Drafts;

- Flexible contracts, which can be adapted to changing environmental needs;

- Regional environmental planning instruments;

- Strategic planning instruments, which should probably be self-standing; their interface to the project level would also require clarification;

- Determination of environmental thresholds;

- An effective inspectorate;

- An adequate appeal process.

6.4.2 Are the Draft Environmental Regulations too Prescriptive or Not?

Participants acknowledged that the Discussion Paper on the Draft EnvRegs is not yet a full working document and consequently it was challenging to answer the question posed in detail. Nevertheless, participants thought the following points should be considered by the ISA as the Draft EnvRegs are advanced:

- While development of the regulations will follow an evolutionary approach, it is important to develop as much content as early as possible. This could include "ramping up" regulatory requirements over time and making these more stringent as appropriate;

- In terms of the principal obligations on contractors, regulatory provisions should be clear and unambiguous. This also includes the roles assigned to the ISA and clarity on the obligations for its respective organs, as well as further refinement on the role of the sponsoring States;

- A prescriptive approach could also reduce litigation;

- Some definitions were considered too prescriptive or in need of further elaboration. Those of "Best Available Scientific Evidence" and "Interested Persons" were considered either too prescriptive or too narrow. The use of "Appropriately Qualified Experts" was considered useful, however, further guidance was deemed to be required with respect to their qualifications, selection process, and the nature and effect of the opinions of such experts;

- "Standardisation" is seen as key to the consistent application of regulatory provisions across the contractor base;

- One observation was made that the current Draft is too long, too detailed, and should be condensed to be attractive to investors. This could include putting some content into annexes to the regulations or in guidelines. Another observation was that the current level of prescription is appropriate, and that further detail should be left for guidelines;

- Defining specific environmental goals and objectives for delivery may reduce the need for defining further details concerning process;

- The review process for the regulations, once adopted, should incorporate specific triggers for review rather than simply being based on time periods.

6.4.3 Gap Analysis of the Overarching Objectives and the Strategic Approach

Overarching objectives were considered to be an essential part of any strategic approach, and largely missing from the Discussion Paper on Draft EnvRegs. Participants acknowledged that the regulations should be based on a global strategic framework for mining in the Area, in order to transparently communicate, in particular, the roles and mandates of the different actors: the ISA and its organs, States, sponsoring States, contractors, and civil society. At the same time, it was proposed that this strategic framework should communicate the overarching strategy for coordinating the ISA's activities and related environmental impact regulation, monitoring, and assessment with the adjacent legal regimes for areas under national jurisdiction and in the water column.

This could be particularly relevant in view of the need for States to establish their own environmental quality objectives and related thresholds in these waters. A strategic approach needs to be reflected on different spatial scales, such as the regional and sub-regional scales, and should also take into account the effects of activities on regions adjacent to the Area into account. Different environmental thresholds may be required for different areas and different resource types, but an overarching approach should be upheld.

As regards to the question 'who is going to do this', there was broad consensus that the ISA is mandated to design a widely encompassing strategy for the Area in relation to mining activities. As the ISA's mandate is limited to mining activities in the Area, more consultation is needed in relation to other sectors and organisations. It was generally agreed that the entire process is of an evolutionary nature in which a 'learning by doing' attitude is appropriate. The comment was made that the approach ought not be too prescriptive and that contractors ought to be given various options and alternatives to reach the set objectives using their own methods.

Environmental objectives were seen as an important tool linking the implementation of the precautionary approach to the commitments of states such as those contained in the CBD and the UN Sustainability Agenda 2030. In this manner, environmental objectives enable management decisions and decision-making to weigh different interests in determining what degree of environmental harm can be acceptable. Overall, a hierarchy of overarching objectives, goals, operational targets, and associated indicators was seen as necessary for breaking down the abstract objectives into measurable and understandable targets, thus avoiding "box-ticking" exercises and effectively aiding the management of activities in the Area. Periodic review should ensure that all objectives are technically feasible.

The overarching objectives should be based on the provisions of UNCLOS and could be set out in the Preamble or as a stand-alone introduction to the Mining Code or its Environmental Regulations. It was suggested that economic and social objectives should be included in addition to environmental objectives.

The following gaps were identified in the Discussion Paper:

- The rights and obligations of the ISA, contractors and sponsoring States are currently not sufficiently distinguished, which may lead to disputes. It is essential that these roles and responsibilities are defined in unambiguous terms in the Regulations. A compliance mechanism for sponsoring States is necessary to prevent the issue of "sponsoring States of convenience";

- Designing a Regional Environmental Management Plan will be a complicated endeavour in regions where effects of activities within national jurisdiction and beyond national jurisdiction overlap. Neglecting effects in areas within national jurisdiction could undermine the purpose of adopting a strategic approach in the first place;

- Some participants felt that "activities in the Area" should be expanded to include transport and processing;

- More focus is needed on the broader environmental context, beyond the mining sites on the seafloor (e.g. climate change);

- It was unclear how to effectively operationalise the precautionary approach;

- Ecosystem functioning is potentially one of the most central considerations when determining environmental impacts but it is not conclusively reflected in the current approach. Ecosystem functioning should feature more prominently in the regulations and specific indicators should be developed;

- Scientific discussion on how to operationalise the term "serious harm" was considered to be necessary;

- The Draft EnvRegs make no reference to the status of the Area and its resources as the common heritage of mankind or its consequences for capacity building, technology transfer, or socio-cultural impacts.

7. Sessions, Presentations and Discussions

7.1 Substantive Criteria

7.1.1 Presentation: Substantive Criteria as Preconditions for the Approval of Exploitation Activity
Robin Warner

Introduction

The ISA is responsible for providing effective protection for the marine environment from the harmful effects of activities in the Area, according to Article 145 of UNCLOS. To meet this challenge, it must determine the relevant environmental governance principles applicable at each stage of an exploration and exploitation activity and how they can be operationalised in practical terms. This presentation discussed some key principles of international environmental law and management which are relevant to the exploitation process and in particular the approval of a Plan of Work (PoW) for exploitation activities.

The questions posed in this presentation were: which generally accepted principles of international environmental law and environmental management apply to activities in the Area and how should they be reflected in the substantive criteria for approval or rejection of an application for an exploitation activity?

The following principles are considered applicable to activities in the Area:

- Common heritage of mankind (CHM)

- Ecosystem approach

- Precautionary principle/approach

- Environmental Impact Assessment

- Best available scientific evidence

- Best environmental practices (including best available technology)

- Transparency – access to environmental information, public participation, and access to justice

- Polluter-pays principle

Options to be deliberated

The CHM principle requires equitable sharing of any benefits from seabed mining in the Area as well as the preservation of the marine environment for present and future generations. To fulfil its obligations under the CHM principle at the PoW stage, the ISA must take into account the financial and technical capabilities of an applicant to carry out successful exploitation activities as well as their ability to ensure effective protection for the marine environment of the Area from the harmful effects of their activities. The applicant must demonstrate its capability to ensure effective protection of the marine environment from harmful effects through processes such as an environmental baseline study, environmental impact assessment, and environmental management plans.

The implementation of an ecosystem approach requires the ISA to assess the impacts of exploitation activities not only on a single mining site/contract area but on the entire ecosystem at regional or sub-regional level through processes such as strategic environmental assessment (SEA) and regional environmental management plans (REMPs). An applicant's environmental baseline study, EIA, and EMP can then be informed and assessed against such an SEA and REMP.

In assessing a PoW for exploitation, the ISA must take into account any uncertainties or inadequacies in the data available, the application of the precautionary approach, and relevant precautionary measures. This entails assessing the capacity of the Applicant to monitor key environmental parameters and ecosystem components so as to identify any adverse effects of mining activities and to provide for the modification of management and operating procedures, as may be necessary in the light of the results of monitoring or increased knowledge of the receiving environment.

In the context of activities in the Area, the requirement to conduct an EIA was recognised as an international legal obligation of sponsoring States and their contractors by the Seabed Disputes Chamber of ITLOS in their Advisory Opinion of February 2011. Some components of the EIA process are reflected in the current Draft EnvRegs. The content of the EIS is not yet prescribed. A notice of the fact that an application for a Plan of Work has been received must be posted on the ISA's website, together with information on how copies of the environmental plans may be accessed. If preparation of environmental plans is discretionary at the application stage, however, this step loses some of its meaning.

The Draft EnvRegs provide the opportunity for "interested persons" to provide input into the EIA but "interested persons" are defined as "a natural or juristic person or an association of persons that in the opinion of the ISA is directly affected by the carrying out of exploitation activities in the Area." This is not fully inclusive of potential stakeholders in the EIA process.

In the context of a PoW for exploitation, both the applicant and the ISA are required to take into account the best scientific information available to them in their evaluation and management of risks to the marine environment. In accordance with this principle, the contractor is obliged to adapt its mining operations to the developing requirements of "best environmental practices" including "best available technology" during the course of its exploitation contract.

A number of international environmental law instruments highlight the importance of access to environmental information and participatory rights in environmental decision-making. The current Draft EnvRegs incorporate some provisions on access to environmental information, consultation in environmental assessment, public notification of exploitation application, and access to environmental plans if the applicant has prepared these.

The current Draft EnvRegs require contractors to bear the cost of pollution control and waste management and give the ISA power to recover costs in the event of remedial measures. The contractor may also be required, if the Council decides in particular circumstances, to post an environmental performance guarantee to secure compliance for remediation, rehabilitation, and removal of equipment from the exploitation site.

Recommendations and Potential Next Steps

The current Draft EnvRegs should be reviewed to assess whether all the international environmental law principles and environmental management approaches discussed above are adequately incorporated.

7.1.2 Plenary Discussion on Substantive Criteria as Preconditions for the Approval of Exploitation Activity

A number of questions related to the principle of the Common Heritage of Mankind (CHM) were raised. One participant requested clarification of the scope of the CHM principle and whether it includes other types of resources, such as genetic resources and ecosystem services in addition to mineral resources. The speaker responded that although CHM also includes the protection of other resources, the equitable sharing of benefits would only apply to mineral resources as defined in UNCLOS. Another participant noted that principles of international environmental law tend to be used as slogans for action rather than understanding their full legal meaning, emphasised that the principle of CHM refers solely to the status of mineral resources and that this restrictive legal definition should be respected. One participant referred to a trusteeship element contained in the CHM principle, which includes the consideration of alternatives and access to justice which derives from the ISA's obligation to act as a trustee. This participant underlined that marine scientific research is also an important benefit to be shared under CHM. The speaker responded in regard to the high standard imposed on the ISA to implement best environmental practice and apply the precautionary approach as elements of a trustee role, but also referred to a limit on the ISA's responsibility due to the definition of "activities in the Area". The duty to cooperate, contained in Article 197 UNCLOS, was mentioned as another guiding value to be considered in regard to DSM.

Another participant inquired whether UNCLOS requires that alternatives to DSM be considered. The speaker responded with reference to the general obligation to protect the marine environment and Article 145 UNCLOS that although DSM and the development of the resources of the Area has been agreed by the States Parties, a balance must nonetheless be struck between conducting DSM and protecting the environment – which could encompass the consideration of alternatives. However, another participant referred to the application of SEA which includes the consideration of alternatives and asked at which strategic level this obligation is activated and who should be the competent authority to evaluate an SEA. The speaker responded that the ISA should be responsible for SEA while the individual EIAs are conducted by the contractors.

The issue of lack of access to justice was raised, to which the speaker replied that notification to stakeholders and the requirement of consultation is part of the EIA process and stakeholder views would then be submitted to the ISA. Aggrieved stakeholders, whether objecting to the approval of a project, or contractors objecting to the rejection of their applications, would need the opportunity to request another hearing at the ISA. One participant noted the existence of ITLOS and its wide jurisdiction over all matters arising from Article XI UNCLOS and asked whether the speaker thought that additional mechanisms for access to justice are needed. The speaker responded that the issue of standing would need to be clarified, and that if aggrieved stakeholders do not have standing, additional mechanisms would be necessary. National processes may provide guidance on how these could be designed. Another participant referred to the closed nature of arbitration proceedings to which NGOs do not have access and noted that the ITLOS rules and procedures do not allow the participation of non-States Parties in proceedings. Mediation and conciliation may also be worth considering as additional mechanisms for providing access to justice.

Another participant asked whether UNCLOS and the Draft EnvRegs contain any provisions as to how a site must be left upon closure. The speaker noted some unclarity about when closure plans are to be submitted but said that provisions for closure plans exist and that the applicant is required to provide more detailed information.

7.1.3   Presentation: Legal Thresholds for Approval or Denial of a Proposed Plan of Work for Exploitation Activity in the Area
Robin Warner

Introduction

The ISA is responsible for providing effective protection for the marine environment from the harmful effects of activities in the Area under Article 145 of UNCLOS. To meet this challenge, it must determine the relevant environmental governance principles applicable at each stage of an exploration and exploitation activity and how they can be operationalised in practical terms. This presentation discussed the potential legal thresholds for approval or denial of a proposed Plan of Work (PoW) for exploitation activity in the Area.

Legal thresholds have mainly been discussed in terms of the point at which precautionary actions need to be taken during exploration/exploitation activities in the Area to prevent harmful effects/serious harm to the marine environment. The precautionary approach comprises three elements: threat of environmental harm, uncertainty, and remedial action. To trigger the obligation to take remedial action, a certain threshold of risk needs to be reached, which can be expressed as gravity multiplied by the probability of harm. Once gravity and probability thresholds are reached, the precautionary approach requires that measures to prevent environmental degradation be taken.

The general threshold for applying precaution is where there is potential for harmful effects to the marine environment from activities in the Area. UNCLOS and the Exploration Regulations set a higher threshold, that of serious harm, for particularly far-reaching measures such as emergency orders to suspend or adjust operations in the Area to prevent serious harm to the marine environment, disapproval of areas for exploitation by contractors in cases where substantial evidence indicates the risk of serious harm to the marine environment, and prescription of provisional measures in a dispute to prevent serious harm to the marine environment.

"Serious harm" is defined in Regulation 1 of the Nodules, Sulphides and Crusts Regulations as: "Any effect from activities in the Area on the marine environment which represents a **significant adverse change** in the marine environment determined according to the rules, regulations and procedures adopted by the ISA on the basis of internationally recognised standards and practices."

In the context of deep seabed mining, ecological thresholds can help to inform the determination of when an adverse change and/or impact may be considered significant, i.e. serious harm. A range of indicators may assist in determining the likelihood of significant adverse changes and impacts at the species, ecosystem, and community levels, including measures of biodiversity, abundance, habitat quality, population connectivity, heterogeneity levels and community productivity.

The threshold for approval or denial of a PoW is not necessarily the same as the threshold for taking precautionary action. The threshold of serious harm is not specifically used in UNCLOS or the current ISA Mining Code in connection with the approval or denial of a PoW for exploration or exploitation activities in the Area. Applying precaution will, of course, be necessary during the course of an application process once the general threshold for precaution is met, i.e. harmful effects to the marine environment. Precautionary actions in the context of the application process involve a series of steps including EIAs, assessment of alternative options, and transparent decision-making.

The application of precaution would not automatically result in rejection of a PoW for exploitation. Amending the PoW and/or changing the location of the mining operation may be an option to lower the risks to an acceptable level. If lowering the risks to an acceptable level is not possible, then rejection of the PoW may be required as a last resort.

Options to be deliberated

A potential threshold that could be applied for determining approval or rejection of a PoW for exploitation is whether the applicant is able to demonstrate an effective system to protect the marine environment from the harmful effects of exploitation activities against a set of objective criteria prescribed by the ISA. Another formulation might be whether the applicant has taken all reasonable steps to demonstrate its ability to provide an effective system to protect the marine environment from the harmful effects of exploitation activities against a set of objective criteria prescribed by the ISA.

In applying this threshold, both the applicant and the ISA would need to engage in a very thorough identification of the risks and uncertainties of the proposed exploitation activities for the marine environment of the Area and the proposed plans to address those risks and uncertainties.

Applicants would need to conduct an EIA and, on the basis of that assessment, develop and submit environmental management plans and a system to address any harmful effects on the marine environment of the Area. The adequacy of the EIA and the environmental management plans would then need to be assessed against a set of objective criteria, approved by the ISA, to measure whether the PoW meets the requirement of effective protection of the marine environment from harmful effects. Some relevant criteria are already included in the Discussion Paper on Draft EnvRegs, in draft regulation 19(2).

Where a risk is identified as a potential significant risk or uncertain risk, further sampling, data collection, and monitoring may be required by the ISA to assess that risk more clearly. Submission of further information and adjustment of environmental management plans may be required of the applicant. Where an applicant is unable to demonstrate an effective system to protect the marine environment from harmful effects after these exchanges and further requests for information, a threshold may ultimately be reached for rejection of the PoW.

Recommendations and potential next steps

Where an applicant for a PoW for exploitation cannot demonstrate an effective plan to address harmful effects on the marine environment against a set of objective criteria prescribed by the ISA, this is then the threshold for requiring amendments so as to meet the required standard or ultimately for rejecting the PoW.

7.1.4 Plenary Discussion on Legal Thresholds for Approval or Denial of a Proposed Plan of Work for Exploitation Activity in the Area

The first topic of the discussion related to issues around determining serious harm. The question was raised whether the threshold of serious harm applies to the entire Area, a region (for example the Clarion-Clipperton Zone), or a mining site, as this has implications for the determination of harm and the amount of available data to determine its significance. The speaker responded that the discussions at the workshop underline that the proposed Plans of Work (PoW) need to be considered from a more regional perspective beyond the actual mine site in order to take into account broader risks. Ideally, an SEA and environmental management plan would be considered by the ISA when examining an individual EIA. It was also asked whether the threshold of serious harm would only apply to vulnerable marine ecosystems (VMEs), to which the speaker responded that flora and fauna should be considered more extensively and well beyond VMEs.

The point was made that there seemed to be general agreement on the definition of serious harm but that natural scientists do not have the necessary data to support the proposed indicators. The speaker responded that this question is a scientific, rather than a legal, question. However, another participant argued that creating thresholds for acceptability is as much a political issue as a scientific issue. The multiple types of thresholds in the Fish Stocks Agreement were referred to as a model for establishing potential thresholds for DSM. The speaker supported this view and agreed that further work is necessary to identify appropriate thresholds.

One comment highlighted that the term "serious harm" is rarely used in UNCLOS, which is a distinct obligation from the duty to protect and preserve the marine environment. The threshold of preventing serious harm only applies for certain procedural cases, such as the exclusion of areas from mining operations or the issuance of emergency responses. Another participant emphasised that the criterion for rejection or approval of an application for a PoW hinges on the applicant demonstrating effective environmental protection from harmful effects (Article 145 UNCLOS) and not in preventing serious harm. In regard to the acceptability of thresholds, one participant noted that it is impossible to determine this without clear objectives for conservation and other objective criteria so that scientists can weigh in on this process. The speaker referred to the SEA process as a vehicle for achieving an overarching environmental vision, which is currently lacking for DSM. More detailed criteria are also necessary to realise this vision.

The need to quantify scientific thresholds in order to inform the technology development process was mentioned. It was highlighted that UNCLOS is one of the few instruments in international law to require sustainable development, which is not intended to prevent an activity from taking place but instead to ensure that when it is conducted, it does not create harmful effects. The technology transfer provisions in UNCLOS may provide an opportunity for the emerging deep seabed mining industry to demonstrate to the international community how technical processes can be designed so that activities can be sustainably conducted.

Another participant responded to the speaker's observation that the application of the precautionary approach is not fully reflected in the requirements of applications contained in the Exploration Regulations. It was underlined that exploration and exploitation exist along a continuum. The speaker responded that the Exploration Regulations do not have the same rigor as the Draft Exploitation Regulations. For example, there is a requirement for conducting a baseline study in both but the Exploration Regulations do not contain a requirement to submit environmental management plans. Precautionary aspects are less apparent in the

Exploration Regulations. Another participant referred to gaps in the Draft Exploitation Regulations, including the definition of specific conservation objectives, the lack of a requirement to identify scientific uncertainties, and a lack of procedural safeguards such as SEAs.

Another participant stated that adaptive management is an on-going process and that important information can be obtained from pilot mining tests and experiences with shallow marine ecosystems and terrestrial mining where there is a precedent for data exchange, collaboration in monitoring, and adjustment to thresholds as changes become necessary. In accordance with this view, it was further mentioned that adaptive management and progressive development of the industry are appropriate responses to the continuing existence of scientific uncertainty, with pilot mining tests being referred to as a critical element in reducing uncertainties.

One participant commented on the issue of economic constraints for conducting baseline studies, to which the speaker responded with the obligation of the ISA to consider the capacity of the applicant to conduct baseline studies as an element of the applicant's overall capacity to implement a PoW when considering whether to approve or reject an application.

Finally, one participant urged that well-developed scientific criteria and overarching environmental objectives are necessary for deliberating on PoW and determining what effective protection means in practice, and asked whether there are examples of such criteria, from other processes. The speaker referred to potential sources of precedent from national jurisdictions for objective criteria such as terrestrial and coastal bioregional planning in national jurisdictions and highlighted the role of regional management plans and SEAs for developing these criteria, perhaps as a parallel process to developing the EnvRegs. A question was then raised whether it would be acceptable to adopt the EnvRegs at all when well-developed scientific criteria and overarching environmental objectives are absent.

7.2 Environmental Standards

7.2.1 Presentation: The Role of and Developing "Environmental Standards" in the Area
Chris Brown and Samantha Smith

Background

In addition to the requirement to adopt appropriate rules, regulations and procedures (RRPs) under Article 45 of the UN Convention on the Law of the Sea, the ISA is obliged to take into account "mining standards and practices, including those relating to operational safety, conservation of the resources and the protection of the marine environment" (Article 17(1)(b)(xii), Annex III). To support and drive the development of appropriate RRPs, the Legal and Technical Commission of the ISA identified a number of high level issues, including two relating to the development of standards[10].

A standard can be considered to be a practice that is widely recognised or employed, especially because of its excellence; it is a way of doing work that is widely accepted as good (or even best) practice.

---

[0]   See ISA Report to Members of the Authority and all stakeholders on a draft framework for the regulation of exploitation activities in the Area, March 2015 (Available at https://www.isa.org.jm/files/documents/EN/Survey/Report-2015.pdf).

Many international standards, including those developed in the offshore oil and gas industry, dredging, trenching and even marine sectors, could be directly applicable to exploitation activities. While there is a likely need to tailor such standards to the specific location and nature of deep-sea mineral exploitation in the Area, there is no need to reinvent the wheel. In addition to those considered by the MIDAS project[11], other examples of leading international standards, which may warrant discussion, include the Global Reporting Initiative, Global Compact, and World Bank Operational Guidelines.

A staged approach to the development of standards is likely. As the industry progresses, industry-driven operational standards will be developed. However, at this stage, the focus should be placed on those standards that are of significance to the respective regulatory bodies, being the ISA and sponsoring States. Equally, the ISA may wish to consider adopting or developing standards that are relevant to its functioning as an international regulator.

Challenges

The extent to which standards will be mandatory or voluntary is a key question. There should be a determination of the appropriate mix of regulatory provisions that prescribe for the adoption of specific technical standards, and standards that seek to operationalise regulatory requirements (e.g. a defined outcome, target, or threshold) by setting out recommended ways of doing things.

An example of an international, process-related standard is ISO 14001:2015, relating to environmental management systems (EMS). Draft regulation 28 of the current Draft EnvRegs uses ISO 14001:2015 principles to set an EMS benchmark rather than prescribing for the adoption of ISO 14001. A further example is that of risk assessment and management. Contractors should have a risk management standard but it is typically not the regulator's job to develop this. That said, a regulation (or guideline) that sets out parameters for an adequate risk assessment and management process that is transparent and auditable would seem appropriate.

One of the principal challenges to the development of specific seabed minerals industry standards is the current immaturity of the industry. The point at which standards should be drawn up will vary as industry development progresses. Unduly prescriptive standards in the short-term may not be appropriate, as they run the risk of stifling innovation. Equally, there will be a need to ensure that there is a regular, periodic revision of standards in response to changes in knowledge and experience. Standard development must also take account of the technical and economic implications of the standards proposed, and potential changes to them, noting that industry needs certainty.

The delivery of an appropriate mix of relevant and feasible standards can only be achieved through an effective and transparent process and procedure for the adoption and development of standards for activities in the Area. The process should be initiated by the ISA and incorporate relevant stakeholders' participation[12].

---

[11]   MIDAS, Deliverable 8.2, *Review of existing protocols and standards applicable to the exploitation of deep-sea mineral resources*, 10 December 2015 at 37-39.

[12]   See *Report of the Chair of the Legal and Technical Commission on the work of the Commission at its session in 2016*, ISBA/22/C/17 at Annex II.

A standard development process

Standards should be developed through a transparent and equitable process. Two key elements of a "good" standard are "content" and "credibility". The content of a standard should be based on good technical information and should be practical and feasible. The credibility of a standard derives from its buy-in by relevant stakeholders.

The starting point for the progressive development of standards for activities in the Area is the drafting of a process and procedure document for the review, development and integration of standards. Such a document could be based on existing codes of practice for the development of standards and should reflect a number of overarching or guiding principles for standard development (e.g. inclusiveness, transparency, effectiveness and relevance, continuous improvement, etc.).

The document should also include details of accountability and ownership of the process (for example the LTC) and the identification of relevant stakeholders involved in the adoption/development of a standard. This should include technical experts, stakeholders who will be required to implement the standard and/or have a direct interest in the adoption or development of relevant standards, i.e. potential contractors (the industry), the ISA, financial institutions, sponsoring States, and environmental NGOs. Consideration must also be given to how such a process will be funded, as the costs and time involved in the standard development process are directly related to the extent of stakeholder inclusion.

The initial process should identify existing and potentially transferable standards (subject to modification, where applicable) and produce an indicative list of standards to be developed across subject areas.

7.2.2   Presentation: Environmental Standards - A First Attempt to Outline the Need
Sabine Christiansen

Introduction and context

The ultimate goal of regulating potentially environmentally harmful activities in the Area is to protect and preserve the marine environment, an obligation of all States set out in Article 192 of UNCLOS. In the Area, this responsibility lies with the ISA, i.e. its member States (Articles 137(2), 152(1)), which shall adopt appropriate rules, regulations, and procedures "to ensure effective protection for the marine environment from harmful effects which may arise from [such] activities" related to seabed mining, including:

- The prevention, reduction and control of pollution and other hazards;

- The prevention of interference with the ecological balance;

- The protection and conservation of the natural resources of the Area and the prevention of damage to the flora and fauna (Article 145).

As the Area and its resources have been designated the common heritage of mankind (Article 136), the ISA bears the responsibility for responsibly developing the Area in the long term on behalf of mankind as a whole. The establishment of standardised procedures, performance, and operational criteria applicable to all actors are potentially the most important steps toward ensuring transparency and efficient regulation in the implementation of Articles 136 and 145 of UNCLOS.

The rules, regulations and procedures adopted by the ISA are binding on all member States, irrespective of their individual consent (Arts. 156–185). The ISA therefore has substantial regulatory power to ensure "uniform application of the highest standards of protection of the marine environment, the safe development of activities in the Area and protection of the common heritage of mankind" as requested in the Advisory Opinion of the Seabed Disputes Chamber of the ITLOS (ITLOS, 2011[13], para. 159). The Advisory Opinion also sets out the most important aims of standardising processes, procedures, criteria and the use of management tools; namely to ensure oversight and control by the regulator in protecting the environment, as well as to establish a level playing field for the operators

Determining the need for standardised approaches

So far, the need to establish standards of different kinds - binding standards, guidelines/guidance, protocols, recommendations, etc. - has not yet been specified with respect to the ISA, as regulator, or with respect to the contractor, in regard to the exploitation phase of deep seabed mining. However, the Working Draft for Exploitation Regulations and the Discussion Paper on the Draft EnvRegs issued by the ISA in 2016 and 2017, respectively, and the set of background discussion documents accompanying the main session presentations of this workshop, provide a wealth of suggestions for topics which warrant the development of a standardised approach.

The implementation of an ecosystem-based approach is suggested as a guiding principle for the management of Activities in the Area, as indicated in the ISA Discussion Paper on the Draft EnvRegs. The ecosystem approach (EAM[14]) has now been recognised in many international fora as the most appropriate way of ensuring the environmentally sustainable human use of particular ocean areas. EAM is based on a long-term vision for the development of a defined region created in dialogue with stakeholders, is integrative, holistic, and based on best available science and knowledge from all sources. As such, it is not sufficient to name the ecosystem approach as a principle in the ISA Environmental Regulations. It, instead, requires operationalisation throughout the Mining Code, including the agreement on operational standards, protocols and guidance. Therefore, the operational elements of an ecosystem approach (vision, principles, operational objectives, assessment, delivery, and implementation) are used here to structure the need for standardising approaches in order to enable EAM in the context of seabed mining in the Area:

- A long-term vision and overarching goals for the future development of the Area and its resources are needed to provide a stakeholder-agreed measure for the acceptability of mining-related activities. Whereas Article 145 of UNCLOS and globally agreed conservation targets will set the framework, the vision and overarching objectives will need to be specified for different regions and time horizons. The vision could be supported by a long-term strategic approach, providing information as to how the ISA intends to implement its mandate and setting the framework for the regulation of environmental matters. A stakeholder-inclusive process under the lead of ISA member States is required, which could also take the form of a Strategic Assessment in parallel with the development of the Mining Code.

---

[13]   ITLOS, 2011. Responsibilities and Obligations of States Sponsoring Persons and Entities with Respect to Activities in the Area, Case No. 17, Advisory Opinion (ITLOS Seabed Disputes Chamber Feb. 1, 2011), at http://www.itlos.org/ [hereinafter Advisory Opinion].

[14]   For the different terminologies that are in use please see: http://www.biodiversitya-z.org/content/ecosystem-approach, for operational guidance see e.g. https://www.cbd.int/doc/publications/ea-text-en.pdf

- Principles act as norm-setting agents governing all activities in the Area. However, the currently proposed principles in the Discussion Paper on the Draft EnvRegs lack the CHM principle, the polluter-pays principle, and transparency in terms of the Aarhus Convention. In particular, the latter requires binding protocols for access and dissemination of information, and a strategy for making best use of scientific and other stakeholder information. All principles require definition and operationalisation protocols.

- The economic, social, and ecological objectives derived from the CHM principle must be implemented using SMART (specific, measurable, acceptable, realistic, and time-bound) management objectives which balance different interests and assess progress towards sustainability.

- Assessment tools range from strategic (Strategic Environmental Assessment, SEA) and regional (Regional Environmental Assessment, REA, see CHAPTER 7.6) to project-specific (Environmental Impact Assessments, EIA, see CHAPTER 7.3). Applied hierarchically, they assist with realising a regulated and sufficiently precautionary approach to determining the acceptability of activities in the Area. SEA and REA are not currently foreseen in the Draft EnvRegs. From the regulatory standpoint, the scope, preconditions, procedures, evaluation, and expected output of SEAs, REAs, and EIAs require the development of a number of protocols. Contractor performance and applicable technical standards require criteria and guidelines, including requirements for environmental baseline studies and data/information delivery. Cumulative assessments and the evaluation of risk assessments and monitoring carried out by the operators depend on standardised data-collection processes. A standard procedure for monitoring and assessment is crucial for enabling meaningful regional compilation of data and information.

- Effective environmental management usually includes a combination of spatial protection measures, such as the APEI network in the Clarion-Clipperton-Zone, and regulatory measures, such as thresholds for emissions and immissions from activities, or environmental effects. For all of these protocols, criteria, and standards need to be set.

- Implementation would benefit from a strategic document, specifying the roles and responsibilities of actors, the use of management instruments, the implementation of adaptive management, and the tools for maintaining oversight and control over activities in the Area, including regional monitoring and assessment plans, and independent research plans.

Next steps

Given the range of standards necessary, a working group including experts from State Parties and ISA observers could determine a priority list of subjects requiring standards, develop a roadmap for delivery, and create specialised expert groups to address particular issues.

7.2.3   Presentation: Risk Management Standards for Regulatory Frameworks and the Ecosystem Approach
Roland Cormier

Introduction

Risk management is a process to ensure that all significant risks are identified, prioritised, and managed effectively. It comprises risk recognition, risk evaluation, risk control, and risk monitoring. A regulatory risk assessment process helps to determine what needs to be done to achieve the environmental and other objectives in a management context. It helps to prioritise the need for measures and to control output via compliance, effectiveness, and appropriateness of measures. One important tool is graphic presentation and categorisation with a so-called Bow-tie diagram, a visualisation of the relationship between events, its causes, measures to limit consequences, and the success of measures as compared to the original objectives.

The Bowtie diagram originated from the petrochemical industries as a health and safety analysis technique in the early 1980s. It was later adopted as an industry standard to manage the hazards related to potential catastrophic events and to provide a systematic approach of assuring control over these hazards. With more than 30 risk assessment tools listed in the IEC/ISO 31010[15,16], the Bowtie analysis is the only controls assessment that analyses and integrates multiple threats and consequences in relation to a central hazard. Following the ISO 31000 risk management process[17], the analysis is used to identify prevention and mitigation measures of a management plan.

What are the management measures and controls needed to prevent and mitigate the cumulative environmental effects generated by deep seabed mining activities?

In this example (Figure 4), the Bowtie diagram sets the source of the risk as deep seabed mining and the event to be avoided as adverse effects to the integrity of the seafloor. The causes are identified as smothering, sealing, changes in siltation, abrasion, and selective extraction of the seabed and subsoil. The detailed consequences are identified as the physical damages having regard to the substrate characteristics and condition of the benthic community. Such a Bowtie diagram is used to identify the measures needed to control the causes of adverse effects to the integrity of the seafloor and the controls needed to mitigate or restore the consequences for habitat and biota as a result of the residual effects of changes to the integrity of the seafloor.

Options to be deliberated

A Bowtie analysis is more effective when conducted within the legislative context of, for example, UNCLOS, and the need to avoid serious harm to the marine environment. The diagram is a representation of the risks of serious harm in relation to the source of the risk (deep seabed mining) and the specific event to avoid (integrity of the seafloor). In this example, the EU Marine Strategy Framework Directive annexes were used as criteria to identify a roster of causes (blue boxes) event (central knot), and consequences (red boxes) as well as the type of controls measures to be considered for the causes and the consequences. Such an

---

[15]   IEC/ISO 31010:2009. Risk Assessment Techniques. International Organization for Standardization.
[16]   ISO GUIDE 73:2009. Risk Management Vocabulary. International Standards Organization.
[17]   ISO 31000:2009. Risk Management Principles and Guidelines. International Standards Organization.

approach could be used to structure an analysis of environmental issues and develop a roster of management requirements for such activity.

The left side of the Bowtie represents the management controls that would be implemented within the operations of a contractor, and can serve as an outline of the environmental impacts to be considered in the environmental assessment. On the other hand, the right side of the Bowtie represents the type of management measures to be implemented to protected ecosystem features and functions at a regional scale. The right side of the Bowtie can serve as an outline of the potential cumulative environmental effects that could occur as a result of this activity and could be used to develop environmental effects monitoring strategies as well as scientific studies.

Recommendations and potential next steps

A Bowtie analysis of a comprehensive set of environmental issues using operational and environmental criteria would deconstruct the current concerns regarding this activity and help identify the key operational activities that are likely to cause impacts. A Bowtie analysis would also help understand what can technically be done to avoid the impacts as well as what would be needed to mitigate the effects on benthic communities. A comprehensive Bowtie analysis of current management approaches for similar activities would provide valuable insight into the design and approaches that could be used for deep seabed mining. Finally, a Bowtie analysis of the current scientific knowledge regarding the causes and consequences of the effects would also identify priority areas for further research and monitoring.

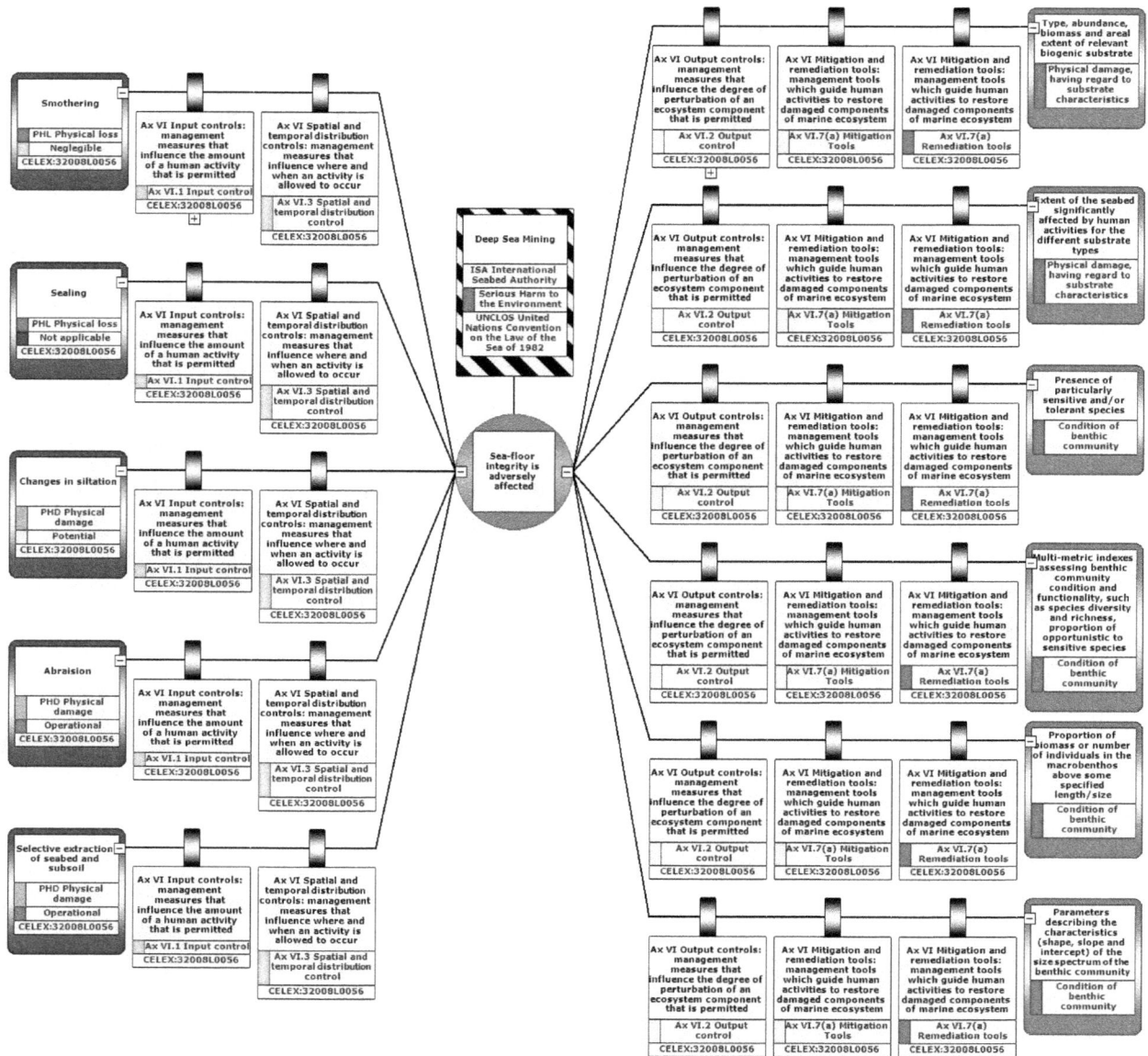

Figure 4: *Risk analysis of operational prevention controls and ecosystem mitigation.*

7.2.4 Working Group Discussions

Develop working methodology to create ISA standards based on existing standards and guidance

The starting point for the development of standards for activities in the Area is the drafting of guidelines / codes for the review, development, and integration of standards. Such a document could be based on existing codes of practice for the development of standards and should reflect a number of overarching or guiding principles including inclusiveness, transparency, effectiveness, relevance, and continuous improvement. The background document "Using and developing standards for activities in the Area: Procedural considerations" (summarised in CHAPTER 7.2.1) was considered a good guide and starting point.

Two topics were discussed in detail:

1.) The process and structure for standard development: who drives the process?

A vision was developed in which the regulator (the ISA and its different organs) should own the process (Secretariat => LTC puts forward standard framework for discussion => Council recommends => Assembly approves). Observer groups, contractors, scientists, international classification agencies, and industry (i.e. various expert fields) could be included in a (sub-) committee on standards under the auspices of the ISA Secretariat which could identify and define the necessary elements of the standards.

2.) Objectives of standards must be pre-defined: why do we need standards (and policies)?

One proposal involved the use of a flow chart to map out the entire application process, from the problem definition stage all the way to the scoping stage for the complete EIS, in order to identify where standards need to be applied and which objectives those standards are intended to pursue. This could be presented as a gap analysis and/or applicability analysis to identify subject areas requiring standards, e.g. resource assessment, biodiversity, and environmental assessment.

The view was articulated that standards are already available that are potentially transferable (e.g. reporting standards).

Use of standards: Compulsory vs. voluntary? Pros and cons

Opinions varied greatly on this topic. Participants first suggested the need to define the terms "standards", "objectives" and "guidelines". There was the view that performance objectives should be compulsory. It was also proposed that guidelines could be compulsory to a degree but flexible in regard to how and where they are implemented. This raised the need to consider what the instrument is intended to do and what it is intended to achieve. The distinction between performance standards and procedural standards was then discussed. Some participants stated that the origins of the standard have bearing on whether it should be compulsory or voluntary. In this regard, a consensus-based standard and a standard developed in another manner would have different needs for anchoring in the regulatory process to ensure that their content is achieved or upheld. Non-compulsory standards raise the issue that the ISA must uniformly apply standards, thus making at least certain elements of their content compulsory.

Another problem which was raised is that standards must evolve quickly and be taken up into regulation in a compulsory form. The development of standards should allow broad participation and should be adaptable to changing needs, which raises the issue how well that can be achieved using a compulsory or a voluntary approach. There appears to be a need to distinguish between the definition of standards being used in the workshop and general technical methodologies, which are also termed "standards". The issue of data interoperability may also require some compulsory elements to make sure that data is consistently usable. It was mentioned that contractors generally advocated the use of voluntary standards to foster innovation. There was acknowledgement that voluntary standards ought to be accompanied with some form of compulsory safeguard, like best available techniques or best environmental practices.

Discussion then continued to whether standards should reflect an achievable minimum standard now or whether they should be aspirational in order to advance environmental protection. The view was held that a mix of voluntary and compulsory elements is likely to be necessary and that the interplay between both forms of standards is likely to be important in the evolution of the regime. Other issues might also figure in, such as the capacity of small island developing States to uphold compulsory standards; although the observation was then made that the requirement of due diligence applies to all actors irrespective of capacity. The view was raised that verification is virtually impossible in regard to deep seabed mining and that there is a large element of trust in the contractors to uphold their responsibilities. The view was also held that voluntary standards are easier for contractors to uphold in their operations. Finally, the participants raised the issue that it must also be considered which standards should be applicable for the ISA in addition to contractors.

Using risk management for the implementation of the ecosystem approach to managing Deep Seabed Mining

In this working group, the concept of a regulatory risk assessment in the context of deep seabed mining was further discussed, and an example process for an IEC/ISO 31010 Bow-tie analysis of a deep seabed mining management plan to address environmental risks was initiated.

The management process starts by establishing the risk management context. In this example, a Bowtie analysis for deep seabed mining sets the risk management context (Figure 5): the source of the risk being deep seabed mining and the event being the integrity of the seafloor being adversely affected. In addition to establishing the authority as ISA, based on UNCLOS, the goal of the risk management approach is to avoid serious harm to the environment. The event (e.g. sea-floor integrity is adversely affected) is an example of an environmental objective that is specific and measurable.

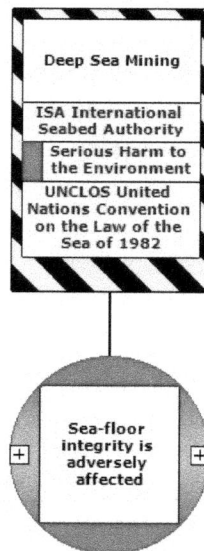

Figure 5:   Risk management context (Bowtie diagram based on BowTieXP).

At the risk identification stage (Figure 6), the causes and consequences of adverse effects to the integrity of the seafloor are listed. The left side of the diagram list the changes that could be generated by the operational activities of deep seabed mining. The right side of the diagram lists the consequences of losing the integrity of the seafloor. For discussion purposes, the causes (blue boxes) would be analysed to identify the relevant changes to the environment resulting from the activities of deep seabed mining. This would subsequently outline the areas that would need operational controls to reduce the likelihood of adversely affecting the seafloor. The consequences would also be analysed in terms of the relevant consequences to seafloor habitat and biota. In this example, the EU Marine Strategy Framework Directive (MSFD) and the criteria for good environmental status are used to create a roster of causes and consequences as an example. Such a roster provides a template to avoid inadvertent omissions when conducting assessments and reviews, while providing transparency as to the potential risks.

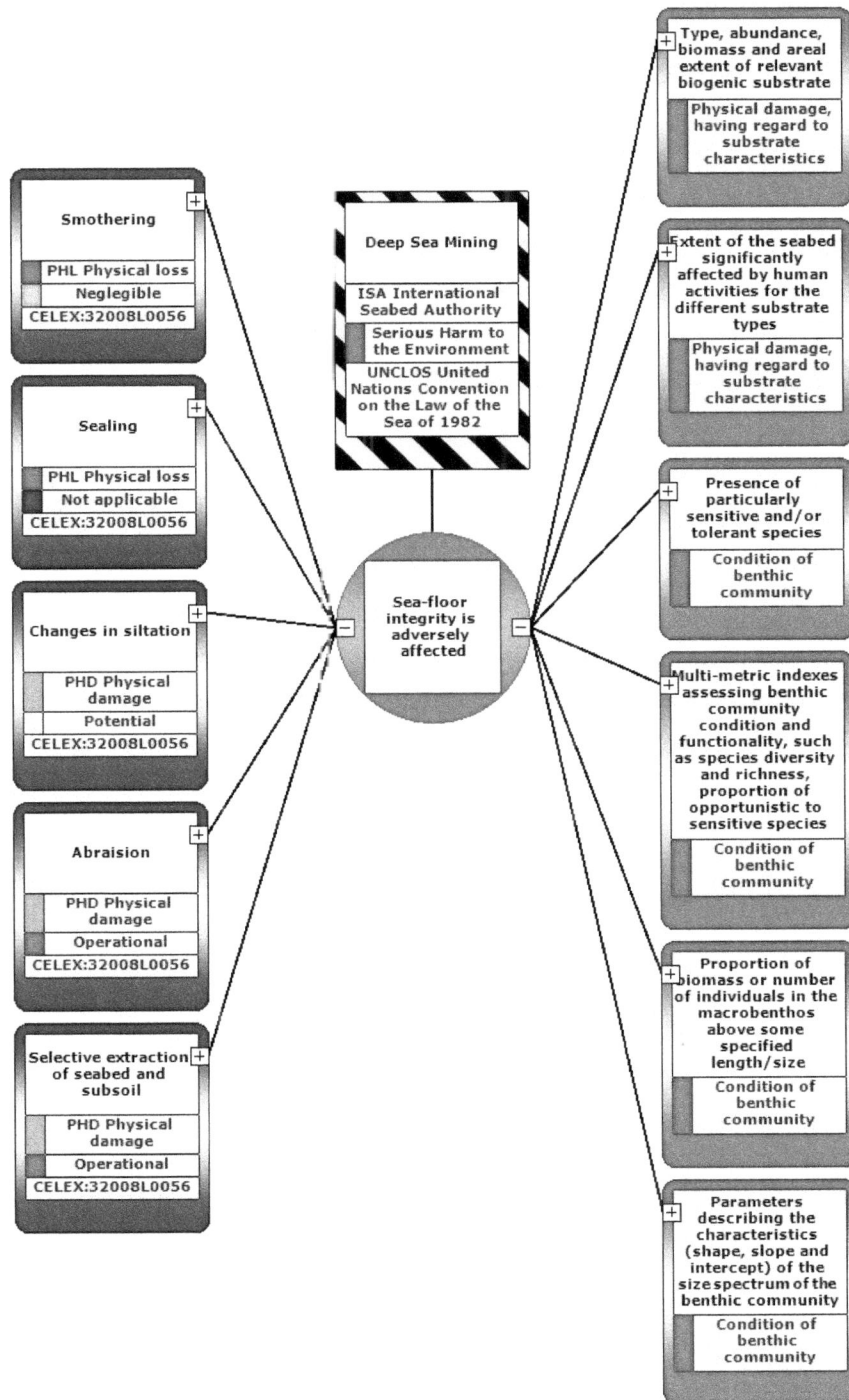

Figure 6: *Risk identification of the causes and consequences of seafloor integrity being adversely affected.*

Operational controls would then be implemented to control the causes and to reduce the likelihood of adverse effects on the seafloor (Figure 6) while ecosystem-based management strategies would be implemented to reduce the spatial and temporal scale of the consequences to seafloor habitats and biota. Input controls and spatial and distribution controls would be implemented for the operational activities while output controls, mitigation and remediation tools would be implemented with respect to consequences. As above, this example uses the program of measures definitions from the MSFD to describe the expected outcomes of the operational controls and ecosystem management measures.

Once completed, the resulting Bowtie diagram is a management plan for a given activity and area.

7.3   Environmental Impact Assessment / Environmental Impact Statement

7.3.1   Presentation: Project-Level Environmental Impact Assessment
Malcolm Clark

Introduction

An Environmental Impact Assessment (EIA) is commonly defined as "the process of identifying, predicting, evaluating and mitigating the biophysical, social, and other relevant effects of development proposals prior to major decisions being taken and commitments made"[18]. An EIA is not a single report, but part of a wider process, with a number of general activities that include: screening to determine if an EIA is required; scoping to identify the issues and impacts; impact analysis to identify and predict effects of the proposal; mitigation and impact management to establish measures to manage impacts; preparation of the report to document all the issues and measures; review process; and decision-making to approve, reject or modify the proposal.

Key Challenges

There is a wealth of international experience with respect to carrying out EIAs[19]. However, whereas the EIA process is well developed in many terrestrial and coastal marine situations and for offshore hydrocarbon resources, guidance for mineral resources is still developing[20].

There are procedural issues with EIA processes, whereby steps to be taken during an EIA process prior to permitting mining tests or operations need to be clearly defined, including the overall scope of the EIA, roles, timelines, scoping procedures, public participation and review, as well as setting performance criteria for environmental reporting and assessment. There are also multiple technical issues which centre on the deep sea being a difficult place to research, and hence obtain robust data on ecosystem structure and function, and limitations of knowledge on the receiving environment flow through into measuring and assessing impacts.

---

[18]   Senécal P., Goldsmith B., Conover S. (1999) Principle of Environmental Impact Assessment Best Practice. pp. 4.
[19]   Glasson J., Therivel R., Chadwick A. (2012) Introduction to environmental impact assessment UCL Press Ltd, University College, London.
[20]   Ellis J.I., Clark M.R., Rouse H.L., Lamarche G. (2017) Environmental management frameworks for offshore mining: insights from experience in New Zealand's Exclusive Economic Zone. Marine Policy in press.

Process issues

- Variable formats and content: A degree of higher-level structural standardisation can make the task of contractors and the reviewing regulatory body much easier.

- Acceptance of an agreed EIA process: This needs to balance existing procedures and regulations of the ISA with accepted international "standards". Scoping for the EIA must occur as early in the process as possible, and should include a preliminary Ecological Risk Assessment (ERA) (or some assessment of risk) to ensure that data collection will support the EIA in focusing on the key elements of impact.

- Role of the ISA Secretariat in reviewing EIAs: The role of the ISA as an involved, regulatory, or facilitatory body needs clarification. It has been suggested that a fundamental change in the ISA will be required for it to undertake a greater regulatory function[21].

- Role of umbrella assessments and plans: Although Strategic Environmental Assessments (SEAs) and regional plans are a relatively new consideration in the context of deep seabed mining in the Area, they are an important context for multiple EIAs that may need to be considered together (see CHAPTER 7.6).

Technical aspects

- Adequate environmental baseline information: Baseline data collection and monitoring studies are important aspects of exploration, as they underpin the preparation of an EIA. Baseline surveys and scientific studies describe the pre-mining state of the environment, as well as some understanding of temporal variability. "Best practice" protocols and standards have been reviewed and updated in recent reports[22,23], and ISA recommendations are currently being considered by the LTC. Test mining has not yet been trialled but is important for improving understanding of impacts as the scale of an operation increases. However, exploration studies will not provide all the solutions to address long-term sustainability of deep-sea ecosystems in the mined region. Knowledge uncertainty will need to flow through into measures to be adopted in the environmental management process (see CHAPTER 7.4.1).

- Requirement for prior ERA: The EIA should focus on the main sources of impact and not spend undue time on elements of little risk. A realistic approach at the beginning of the exploration phase is to conduct a qualitative assessment to guide data collection and support a subsequent, more quantitative assessment to evaluate activities identified as high risk.

---

[2]   ISA Technical Study No. 16, Environmental Assessment and Management for Exploitation of Minerals in the Area. Report of an International Workshop convened by the Griffith University Law School in collaboration with the International Seabed Authority in Queensland, Australia, 23 – 26 May 2016. ISA (2017)

[22] Billett D., Jones D., Murphy K., Gjerde K., Calaco A., Morato T., Cuvelier D., Vercruijsse P., Rolin J.F., Ortega A. (2015) Review of existing protocols and standards applicable to the exploitation of deep-sea mineral resources. Deliverable 8.2 of MIDS project: pp. 148.

[23] Swaddling A., Clark M.R., Bourrel M., Lily H., Lamarche G., Hickey C., Rouse H., Nodder S., Rickard G., Sutton P., Wysoczanski R. (2016) Pacific-ACP States regional scientific research guidelines for deep sea minerals, Noumea: pp. 123.

- Desired formal structure of the EIA/EIS report: This can build on input from several reports and draft templates[21,24,25] and can be finalised shortly after the workshop.

- Acceptable content of an EIA: There has been a developing trend to widen the scope of EIAs to a "whole of the environment" approach, with a balanced consideration of both biophysical and socio-economic impacts. Draft templates currently include "social impact assessment" as well as consideration of cultural and economic factors, although these are less developed and extensive than environmental impacts. These additional considerations may be less important in the Area than in national situations, but should be considered. The variability in environmental characteristics between resource types and locations means a careful balance is required between EIA guidelines being highly prescriptive and being too general (where adequate standards are not clear), which is an issue addressed by Clark et al. (2017)[24] who provide more explanation on what should be included under the template headings.

Key recommendations and options

- Definition of an agreed EIA process that is consistent with ISA responsibilities and can be incorporated into the developing Exploitation Regulations. This may include revision of the process where it links with existing Exploration Regulations (e.g. inclusion of an initial scoping report and risk assessment). It can be developed from options given in ISA (2017)[21];

- Clarity on what an EIA contains in terms of its general scope. Should the EIA extend to include a general account of social, economic, and cultural aspects where appropriate?

- Agreement on an EIA/EIS template (to match defined scope);

- Clarification of roles of various end-users and interested parties in the EIA process, especially public engagement and review processes;

- The evaluation of an EIA should link with issues on how residual risks and general scientific uncertainty are managed in the environmental management plan process;

- Definition of the general nature of baseline data required during exploration that will support the EIA. This can draw on recommendations from ISA, SPC-NIWA, and MIDAS reports;

- Specification of the general nature of a standard monitoring programme for test mining and exploitation phases. This links with baseline data requirements, defining key environmental indicators to measure, and consideration of frequency and duration.

Further workshops to consider the role of Impact Reference and Preservation Reference Zones, standardisation issues, and the development of standard baseline and environmental monitoring schemes may

---

Clark M.R., Rouse H.L., Lamarche G., Ellis J.I., Hickey C. (2017) Preparing Environmental Impact Assessments: General guidelines for offshore mining and drilling with particular reference to New Zealand, NIWA Science and Technology Series [https://www.niwa.co.nz/coasts-and-oceans/research-projects/enabling-management-of-offshore-mining]: pp. 107.

25   Swaddling A. (2016) Pacific-ACP States regional environmental management framework for deep sea minerals exploration and exploitation, Noumea: pp. 100.

be needed to determine what is required to support a full "ecosystem approach" and move more towards ecosystem structure and function than partial community descriptions.

7.3.2 Working Group Discussions

Finalisation of EIA/EIS templates

The working group made four general observations about the current ISA draft EIA template, which would improve the final document. It first observed that the current template places too much focus on national issues rather than on those issues falling under the ISA's mandate. For example, while communication and stakeholder engagement were considered an important part of the EIA process at the Gold Coast Workshop[26], it was pointed out that this is much more an obligation of sponsoring States than the ISA. This issue could be addressed by "denationalising" some issues in the EIA process. The group was not able to resolve the issue that the EIA process is confined to the Area due to the ISA's mandate and does not represent a "cradle-to-grave" approach.

While the group acknowledged that the ISA does not have control over several relevant aspects of the technical process, it nonetheless felt it appropriate that the EIA indicate that environmental standards will be upheld in all elements of the process. It was then observed that the current template is not tailored to specific deep-sea habitats. Although it addressed benthic invertebrates, it failed to include protists and microbes which play important roles in the deep-sea environment. It was agreed that the template should continue to examine individual taxa but that additional guidance for contractors should also be included to ensure that community linkages, connectivity, trophic dynamics and ecosystem modelling are discussed in an additional section. Third, the group found that the template's approach is too descriptive, rather than focusing on what the EIA needs to produce as a final product in order to effectively inform environmental management and monitoring plans. Addressing this issue would essentially involve working backward to determine the necessary content to achieve the EIA's objectives. It was also observed that the current version of the template does not reflect the ecosystem approach but is instead much more driven by potential impacts on biological communities. Finally, the template headings need to be revised to ensure that the terminology is consistent with that used in the future Environmental Regulations and LTC recommendations.

EIA process: Roles and responsibilities

Discussion in this working group addressed central questions concerning the obligations of different actors in the EIA process: who does what, who must be consulted, who reviews, and who ultimately decides. Regarding "who", contractors, various bodies within the ISA, sponsoring States, members of the public, interested persons, experts, and different types of observers were all identified.

A point was made that all EIA processes would be accompanied by iterative discussions.

Before addressing scoping and screening, the discussion first considered the organisation of baseline studies. It was recommended that these should be prepared by the contractor with guidance from the LTC, reviewed

---

[26] ISA Technical Study No. 16, Environmental Assessment and Management for Exploitation of Minerals in the Area. Report of an International Workshop convened by the Griffith University Law School in collaboration with the International Seabed Authority in Queensland, Australia, 23 – 26 May 2016. ISA (2017)

annually and that contractors should receive feedback and review by the LTC. Such baseline studies should be guided by the terms of the contract and the Exploration Regulations.

The multiple uses of data were also discussed, including their role in informing EIA and SEA processes and developing regional environmental management plans, as well as how contractors might use data to justify taking particular approaches. Some participants emphasised the need to use structures already set up by the ISA and potentially to set up an environmental committee to review baseline data and ensure better consistency between contractors.

Different views were expressed on the issue of open access to environmental information and reference was made to steps currently being undertaking by the ISA towards greater accessibility. Some participants proposed that the structure of the LTC needs to be changed, perhaps by creating a subsidiary body with the purpose of increasing the level of expertise available to the LTC to support its decision-making process. Participants recommended considering best practices on the use of expert advice in other regimes.

Another point was made that the feedback process between the LTC and the contractor needs to be regularised, and that the Secretariat could play a mediating role as it is not part of the final decision-making process.

Discussion then considered whether the scoping study should be made available for public consultation. The process of public consultation in New Zealand was referred to, where, although numerous replies were made during public consultation, only a few of these actually required a response from the decision-maker.

Finally, a point was made that pilot mining tests will require their own separate EIAs and will be a good opportunity to test the adequacy of baseline data before they become the basis for a full EIS and ultimately for a commercial mining operation.

Standardisation of assessments and monitoring

Discussion in this working group focused on the issue of scoping, beginning with a consideration of the procedures used in the German context in regard to offshore windfarms in the German EEZ. This opened the discussion as to how practices in other sectors can be taken up in DSM. A number of questions arose as to whether scoping is mandatory or optional in applications for a Plan of Work, should it be required, and what the consequences would be of both approaches. Discussion then considered what purposes the scoping exercise might serve. Examples given by participants included setting out expectations for the EIA process and giving security to the contractor prior to making investments in technologies and environmental management. Discussion then considered how much iteration is involved in the scoping process, the degree to which it should be prescriptive, and who should be involved in an iterative process beyond the applicant and regulator, making reference to the issue of public participation. The point was made that a successful scoping exercise requires a mix of prescriptive and iterative elements and should take a site-specific approach tailored to the claim, with the flexibility to take into account what might be necessary in a specific situation, while nonetheless addressing a general catalogue of issues common to all scoping studies to ensure quality and comparability among all applicants. Prescriptive elements were considered to provide clarity to applicants and scientists regarding their roles and responsibilities as well as identify, the respective time periods for collecting the required information. Finally, working group participants discussed how best to integrate scientific information into the scoping process, and it was proposed to establish standards for the variables in a scoping exercise, that allow changes to these variables as further information becomes available.

7.3.3 World Café Discussions

When are environmental impacts significant, how can these be determined and which consequences do these have?

The severity of environmental impacts is context-dependent and some participants felt that the significance of impacts is not only a scientific question but also involves societal value judgement. Summarising the views of different participants, impacts were considered to be significant when they:

- Violate against UNCLOS, i.e. cause marine environmental pollution or significant adverse change in ecosystem diversity, productivity and stability of biological communities or VMEs, or when they interfere with the ecological balance of the marine environment (Article 145);
- Are irreversible and/or not mitigatable (e.g. extinction of species, loss of keystone species or processes);
- Fall outside of natural variability;
- Cause changes or losses in ecosystem functions;
- Affect unique or rare communities, and threaten their existence;
- Are contrary to predefined environmental objectives, i.e. reach tipping points or some predetermined thresholds (e.g. using operational indices such as a standard matrix of impact severity or multi-dimensional matrices of ocean health in EIAs);
- Cause persistent loss of, for example, > 5% of the biodiversity of the relevant ecosystem.

The question of how significant impacts can be determined revealed various degrees of interpretation, ranging from the natural scientist (statistical) perspective on how to identify environmental change to a higher-level operational perspective on impact severity:

- Comparison between Impact and Preservation Reference Areas at higher taxonomic levels (IRA and PRAs have to be delineated through spatial planning and should be similar in nodule abundance and species assemblages);
- Comparison with baseline conditions using similar methodology and various statistical tools;
- Comparison with reference subsystems or indicator species in other systems, e.g. by applying BASE and PA;
- Comparison with pre-defined standards and thresholds, as stated in the EIA/REMP;
- Physiologically-based ecosystem modelling to determine priority processes and species;
- By using ecosystem criteria (loss of function);
- Multi-scale analysis: cumulative determination of the spatial and temporal scale of impacts with respect to area, diversity, abundance, and distribution (statistical significance requires reasonable sample sizes and defensible numbers);
- Through value judgement (social/political dimension): determination of acceptability.

The world café groups took different approaches as to what the <u>consequences of significant impacts</u> could be. From a procedural point of view, there is an obligation to manage activities in order to prevent serious harm – with serious harm being a ground for refusing an application or setting it back for further work in terms of mitigation. At the proposal stage and at the operational stage, the ISA retains the capacity and the legal mandate to use emergency orders (Articles 162 (2) (w) and 165 (2)(k) of UNCLOS), which can be applied to modify, stop or suspend operations which could include operational adaptations as the activity is under way.

From a scientific point of view, exceeding the tipping points in biological and ecological systems will lead, amongst others, to loss of biodiversity, ecosystems services, ecosystem processes and functions (taking note of cumulative effects), loss of scientific or economic values, and loss of the value of the heritage of humankind.

In terms of the responsibility and liability of the contractor and the sponsoring State to <u>avoid</u> significant harm, some measures could be:

- To install additional safeguards from the ecological and genetic point of view (e.g. create a genetic repository for deep-sea species);
- To improve definitions so as to reflect the full range of knowledge of consequences: consider mitigation measures, stimulate research on community recolonisation, expand baseline studies to the regional scale beyond the impacted/mining areas, and improve spatial planning procedures to conserve ecosystems.

Some of these things can be done before contractors are allocated. Others need to be done later. One suggestion was to declare the whole seabed a Marine Protected Area and allow mining to proceed only in those areas where the applicant can demonstrate that serious harm will not occur, e.g. that all species in the proposed mining area exist elsewhere.

How to deal with uncertainties

As the deep sea is a remote place with comparatively poorly-studied ecosystems, and deep seabed mining is a hitherto unproven and untested activity, our understanding of the potential impacts of the activity on deep-sea ecosystems is hampered by a high degree of uncertainty, in particular with regards to the recipient environment and options for mitigation through technological optimisation. The discussions on how to deal with uncertainties were streamlined into two different categories:

Management of uncertainties in scientific/environmental data and knowledge:

- Environmental baseline data collected by contractors should be made available to other contractors, scientists, the public, etc.;

- A gap analysis using all available environmental data would be helpful to identify uncertainties and consequently to focus research activities on assessing those gaps;

- There is a need for standardisation of data collection;

- The precautionary principle should be implemented through adaptive management (e.g. see Articles 6.2 & 6.3 of UNFSA for potentially useful legal guidance) / disclosure / transparency;

- Uncertainty must be quantified using statistics;

- There is a need to decide upon precautionary thresholds;

- The acceptable level of uncertainty must be determined;

- Long-term and large-scale observations (e.g. of natural variability) are needed;

- Indicator species to check the status of an ecosystem should be identified from baseline studies and used during monitoring purposes;

- Insufficient monitoring leads to a lack of data acquisition. This may lead to suspension of the mining activity in order to revise the operational plan for mining;

- There is a need to define priority actions during monitoring.

Management of uncertainties concerning technology:

Uncertainties arise as different mining technologies and their expected impacts on the environment have not yet been sufficiently tested. The following measures were proposed to reduce uncertainties in this regard:

- Use of BAT/BEP: independent expert review could be helpful. Some participants felt that confidentiality until exploitation (e.g. testing of mining and monitoring technologies, technologies for propose of restoration) is essential;

- Development of technologies must integrate the principle of impact minimisation through adaptive development, including reduction of the sediment plume. New technologies are required to cope with this principle. Sensor CCTV must be continuously developed;

- Obligatory reporting of environmental impacts caused by tests of each mining component is required.

Consequences of the EIA for decision-making

It was acknowledged that consequences from EIAs arise for the ISA, contractors and sponsoring States. The topic was split into several more targeted questions for clarity.

What does approval of an EIA actually mean, i.e. what is being approved? Two different interpretations of what an EIA process achieves came up: 1) An EIA is approved when the impacts remaining after mitigation are deemed to be acceptable, and 2) An EIA provides a fair assessment of impacts, but does not postulate whether these impacts are acceptable or not. Consequently, the question arose as to who will determine whether risks are acceptable or not, and when will this occur during the decision-making process? Would this be an internal or external process? The current decision-making processes of the ISA appear unsuitable for handling EIAs in detail but it was unclear what the alternative could be. However, an increase in the capacity of the ISA Secretariat, in particular also with regards to an inspectorate, was considered urgent, in order to enable year-round work by a team of professional experts. The team should include marine scientists from different disciplines, eco-toxicologists, and geochemists. A review committee was proposed. There was a call for the establishment of new institutions.

Furthermore, the meaning of a final decision on an EIA (yes, no, maybe) is procedurally unclear and needs clarification, e.g. as to options for revision and appeal. A "maybe" could be accompanied by a request for more information on, for example, extended monitoring requirements or more sampling to demonstrate the kinds of mitigation measures that might be undertaken. Two different opinions on how a positive decision on the EIA approval process might influence the overall Plan of Work (PoW) for exploitation were expressed:

- The EIA is just one element of an approval process (other elements are, for example, the demonstration of financial and technical capability, a feasibility study, a closure plan). However, it should be given priority in the overall decision on whether to proceed or not because it either affects or is affected by other decisions; or

- All elements of an approval process should be equally important in determining whether or not to approve a PoW for exploitation.

For the ISA, results from project-specific EIAs can and should feed into regional environmental management plans.

Regarding the decisions to be made by the contractor, an EIA affects both upstream and downstream requirements and decision-making, i.e. the thorough collection of environmental baseline data, preparation of EIA documents, choice of technology, location(s) of the IRZs and PRZs, decisions on whether to proceed or not in the process, the level of insurance, contribution to environmental bonds, and decisions relating to the environmental management plan and the PoW for exploitation. Each step of the EIA-process from scoping to screening to the assessment itself is a point of decision.

The EIA is an extremely important part of the PoW. For an EIA to be effective, much greater access to information and access to justice will be necessary. Noting that the ISA is working towards increasing access to information, it was noted that there may be a conflict of interest between proprietary technical information and public environmental information needs. The roles of the public, interested persons, and appropriately qualified experts, and when during the EIA process they will have opportunity to participate, still needs clarification, as well as how these terms will be defined and these people identified. In the case of conflicts, the last resort could be the International Tribunal for the Law of the Sea.

A suggestion was made that the ISA could stimulate the formation of contractor consortia with different strengths, for example enhancing the cooperation between a contractor with very strong financial capabilities but for whom environmental data and analyses are lacking, and a contractor with better potential for good EIA results.

Several additional instruments were proposed that could usefully be considered in the drafting of the Exploitation Regulations:

- An iterative review process as part of the EIA procedure (if rejected, a modified application/re-application should be possible);

- Time limits for decisions during the whole application process for all actors;

- If rejected, a "conciliation" process (by expert opinions?) should be possible before seeking legal redress;

- Financial guarantees should be required for the case that a future contractor fails to comply with the conditions set by the EIA.

7.4   Adaptive Governance

7.4.1   Presentation: Adaptive Management
Neil Craik

Introduction

The presence of scientific uncertainty in relation to the ecological impacts of deep seabed mining has led to increased interest in the use of adaptive management (AM) tools as part of the environmental regulatory structure for the exploitation regime. AM is a form of structured decision-making that addresses the uncertainty of prediction with respect to the efficacy of environmental management interventions by monitoring of the effects of the management plan and assessing the results of the monitoring with the intention to learn from the results and incorporate findings into revised models for management actions. While widely acknowledged as a useful management tool, careful consideration must be given to the bio-physical and management conditions under which AM processes are implemented.

Challenges and Key Issues

In considering the development of AM practices for activities in the Area, several key challenges can be identified:

1.) Is the deep seabed mining regime suitable for the use of AM? In particular,

- Is there sufficient scientific certainty around key environmental parameters to avoid irreversible or catastrophic risks?

- Does the legal and operational framework allow for adjustments that are sufficiently controllable?

While AM is a response to uncertainty, there must be reasonable levels of understanding of the environmental conditions to ensure that activities are not approved with undue reliance on the future ability to respond to unanticipated environmental effects. Consideration needs to be given to the levers that are available as mechanisms for adjustment, such as increased mitigation measures or adjustments to operational activities. It remains unclear at this stage what environmental metrics may be measurable and whether there are suitable management activities short of stopping the mining activity that would ameliorate unanticipated adverse impacts.

2.) Should the adaptive management approach be primarily directed towards management activities, with less formal emphasis on testing specific scientific hypotheses?

AM in practice has a high degree of variability in its form and in particular in the extent of its adherence to a highly structured experimental process. Defining adaptive management too loosely and without reference to explicit scientific goals and methods may lead to treating adaptive management as a form of contingency planning and "trial and error" environmental management. On the other hand, the resource and capacity requirements for a strongly inquiry-oriented approach may be higher, and may be less responsive to the operational requirements of contractors.

3.) Should adaptive management processes be directed at environmental management plans only or should adaptive management be applied to higher order policy and regulatory instruments?

The background reports and the structure of the Draft EnvRegs suggest a focus on project- specific adaptive management. The vehicle for implementing AM will depend on the final structure of the environmental approvals but the current direction is that each project will require an environmental management and monitoring plan, and that this instrument could be subject to AM processes. It is possible for AM to be applied to higher order instruments, such as regional plans, although other tools, such as periodic reviews, may be more suitable to non-project specific interventions.

4.) Does the legal framework surrounding security of tenure constrain adaptive management approaches?

Under the deep seabed regime, security of tenure of contracts is guaranteed under UNCLOS and standard contract terms, which requires careful attention to drafting to ensure that the imposition of environmentally necessary AM measures do not conflict with security of tenure guarantees. Balancing the need for investment certainty with responsive environmental measures raises further issues regarding the procedural rights afforded to contractors and other stakeholders where AM procedures require significant operational adjustments.

Next Steps

The consideration of AM as a constituent part of the deep seabed mining activities requires further clarity on the potential environmental metrics subject to monitoring that might trigger AM, and the range of interventions that would meet the requirements of environmental harm avoidance and investment certainty. It may be possible to identify a set of principles that would govern AM plans, including the instruments to which they attach, and the minimum components of an acceptable AM plan, but leaving the specifics of the AM plan to be developed in the context of specific project approvals. A second useful step would be to identify those elements of the deep seabed mining regime that may conflict with the implementation of AM, with a view to addressing any interpretive ambiguities.

7.4.2   Presentation: Adaptive Management with a Strict Approach and a Broader Concept
Guifang Xue

The concept of adaptive management (AM) emerged in recent years along with the development of deep seabed mining (DSM) regulations under the mandate of the ISA. It is deemed to be an important management tool for robust decision-making, to deal with scientific uncertainty in the process of regulating exploration and exploitation activities for the mineral resources of the Area. Nevertheless, the operationalisation of AM should be prudently applied with a strict approach but a broad concept.

The Terms

The AM approach reflects a learning-by-doing process, but is not a panacea with infinite and inscrutable flexibility. Given that the contractors (applicants/miners) require certainty due to the large capital investment required for the proposed mining operations, adaptation and adjustment of operational rules and standards must be based on the prerequisite of security of tenure and legal certainty, with clear and feasible management goals. The application of AM should be taken with careful selection of the items to avoid unnecessary suspicion and scepticism throughout the DSM regimes.

The effective application of a strict approach also needs to promote a broader concept to encourage the collective participation of all actors. In particular, the crucial role of the contractor for the development of a regulatory framework cannot be neglected. AM cannot compensate the lack of baseline environmental data or inadequate modelling results. Contractors, as main operators in collecting baseline data, are to be encouraged to collect the data as accurately and comprehensively as possible, both inside and outside of their contract areas, such as in the APEIs. Paucity of data in the Area is detrimental to better understanding the marine environment and possible mitigation of anthropogenic impacts resulting from DSM activities.

The AM approach emphasises a structured decision-making and learning process from the management perspective. For regulators, caution must be taken to avoid "side effects" of AM by combining strict rules with general principles. Specifically, detailed rules/conditions together with specific requirements and procedures need to be clarified to ensure certainty and improved management outcomes.

The AM concept also emphasises communication and engagement by applying AM to a broader range of actors related to DSM, particularly contractors to ensure a "two-way traffic" towards the same destination. Efforts are required to involve contractors by stimulating their "self-control" incentives and desire/expectation for a "better" outcome. The ultimate goal is to achieve a collaborative and effective partnership by sharing the management and monitoring burden.

Application

Effective application of a structured decision-making and learning AM process entails three basic elements:

- Clearly defined, appropriate and feasible objectives;

- Sensible and reliable understanding of the system and its operation; and

- Capability to ensure the smooth and effective operation of the system, enabling adjustment towards the objectives.

When applying this procedure, issues need to be defined from both the regulator and contractor perspectives, the objectives to be identified taking into consideration the interaction between policy, science, technology and evaluation criteria, and being formulated with agreed triggers in an objective and transparent decision-making process.

The application of the AM concept needs to engage a broader range of actors, particularly contractors from the R&D phase to the exploitation stage, including their exploration plans and pilot mining tests at gradually increasing water depths (e.g. 1000, 3000, 6000 m) and scales. Contractors are encouraged to pay close attention to the monitoring and assessment of environmental impacts, to developing and improving monitoring methods and skills to reduce the disturbance to marine environment, and to the design of mining systems using best available science and technology. Contractors may also be encouraged to set the objectives to minimise and reduce the disturbance and damage to the environment, to allow adjustment and adaptation to critical parameters and structures, and to use new and environmentally-friendly technologies.

7.4.3   Working Group Discussions

Is the deep seabed mining regime suitable for the use of adaptive governance and management?

The working group was of the view that Adaptive Management (AM) should be applied in the regulation and management of deep seabed mining, and should include stakeholder engagement. In addition, the following questions were raised concerning the implementation of AM in practice:

- Can management objectives be set explicitly for AM? Yes.

- Can resource relationships and management impacts be represented in models? Yes but it should be noted that some effects only become visible over very long time scales, which makes a "learning-by-doing" approach inappropriate. Significant uncertainties will remain, and monitoring, in particular, will be an essential tool for applying AM.

- Is AM realistic? Yes but it depends on the prerequisites for the process, such as thresholds. These need to be agreed on in advance, as well as how corrective action will be taken.

- Could a pilot mining test or staged mining be a form of AM? Yes but scaling up may not be the situation found with deep seabed mining. Technology decisions and adaptation of technologies are crucial elements of AM, however.

- AM - yes or no? Yes but AM cannot be seen as a replacement for an initial set of rules and regulations. Participants raised the concern that AM may emphasise short-term effects, and fail to take longer-term effects into account. Monitoring and controllability will be important for implementing AM in practice.

- How should economic and environmental factors be weighed in the decision-making process? Economic feasibility must operate within the boundaries set by environmental objectives. There is a need for indicators that can be used for monitoring the changing state of the environment. As many of the effects will be long-term, appropriate proxies are required.

Options for an adaptive regulatory framework to be incorporated into the regulations

The topic of this working group was expanded to include whether an adaptive regulatory framework should be incorporated into the regulations themselves, into operating procedures, as separate recommendations concerning REMPs, EIAs, and EMMPs, or as part of the review of Plans of Work. The Group was of the view that mining should not be allowed to proceed to the point of causing or threatening to cause serious harm. One proposal involved incorporating the thresholds into the regulatory process so that a regulatory response would be triggered prior to a threshold being reached and well before a threat of serious harm has become apparent. This option could also be incorporated into EMMPs. It was argued that contractors would be in a better position than the ISA to apply adaptive management in this context. This led to the opinion that this option should not be overly prescriptive in the regulations but should instead allow some flexibility at the operational level.

The participants agreed that the regulations must ensure legal certainty, likely through guidelines, templates, and trigger thresholds. In unforeseen conditions where triggers and corresponding actions cannot be specifically defined, the EMMP could include a discretionary clause authorising the inspectorate to order further actions when deemed necessary. The Draft Exploitation Regulations establish the "threat of serious harm" as the legal threshold requiring the ISA to take emergency measures. In the event of a dispute whether or not actions ordered by the inspectorate are mandatory, the inspectorate's orders should be implemented until an agreement has been reached using an appropriate dispute resolution mechanism. Participants also proposed the periodic review of regulations, guidelines, and other documents using existing mechanisms. It was mentioned that all options require effective monitoring technologies and an obligation to immediately respond. It was further discussed that REMPs should also be adaptive. In order to ensure an effective AM system, contracts must include references to the REMP and EMMP, and require contractor compliance. Breach of these contract provisions would constitute grounds for suspension of the contract.

Opportunities and obligations of contractors to adapt their mining operations after the contract is concluded (e.g. through innovation / development of BAT and BEP)

Fruitful discussions among the participants of the working group resulted in the following 11 statements on options for contractors to adapt their mining operations after the contract with the ISA has been concluded:

- Adaptive Management (AM) during operations is possible and in the dredging industry, though in another environmental context, some operators like DEME do it all the time[27].

- In order to allow the contractor to be innovative and work towards the best result, it is necessary to achieve a balance between prescription and flexibility in the contract and the regulations. Finding this balance is a challenge.

- It was suggested that an important step for developing the AM parts of the contract would be to include a mitigation plan in the Plan of Work (derived from the EIA). A review of the definition of mitigation would be useful.

- Security of tenure aspects in developing post-contractual AM regulations, and BAT/BEP requirements, including a definition and function of "material change", must be taken into account.

- Developing incentives to encourage the use of, and develop cost-effective approaches to AM and BAT/BEP should be explored (see also the working group discussion on incentives in CHAPTER 7.5.2).

- It is important to consider that AM is also influenced by regulations from other intergovernmental organisations (e.g. IMO, LC/LP).

- AM and developing criteria for suspension of operations: if AM is not enough to prevent serious harm, it must be considered what the criteria to determine serious harm are (still needs definition), and how these could trigger suspension as the only, albeit ultimate, response. Practical questions of

---

27   Example: Ecoplume: operational proactive environmental management of dredging.
    https://www.dredging.org/digitallibrary/Abstract.asp?menu=&v0=1026

implementation must also be considered. For example, who monitors this, and how? The only source of information is from the contractor who is conducting the operations.

- The implementation of AM and the concept itself will change as knowledge and experience are gained. Therefore, the exploitation contracts with the ISA will also evolve accordingly, and the first exploitation contracts should be the most flexible in AM terms. However, there are issues on "equal treatment" (required by UNCLOS) that must be considered with this approach.

- A system for comprehensive incentives is necessary to ensure a continuously evolving, level playing field.

- Regulations must be clear on who revises, when to revise, and how to revise contracts to address AM responses once contracts are concluded and are being implemented.

- AM also has implications for sponsoring States and these must be considered.

7.5   Pilot Mining Tests[28]

7.5.1   Presentation: Pilot Mining Tests - Legal and Regulatory Issues
Katherine Houghton

Background

Technologies for deep seabed mining are rapidly advancing. It is logical that developers wish to conduct tests on equipment and systems to determine whether they can eventually be used to commercially extract minerals from the Area. Although UNCLOS contains provisions for testing small, medium, and large-scale mining equipment in the Area under both exploration and exploitation licenses, rules and procedures for conducting pilot mining tests (PMT) are currently unclear and sorely inadequate. This was recently acknowledged during the initial drafting process for the Draft Regulations (July 2016), particularly with regard to Environmental Impact Assessment (EIA) requirements prior to test mining (Commentary, Section 2, 31). This is despite the clear understanding among stakeholders that the results of PMTs are critical – likely the "primary inputs" – in determining the potential environmental impacts of commercial-scale mining. The further development of both the Draft EnvRegs and the Working Draft for Exploitation Regulations is therefore an important opportunity to define an appropriate process for regulating PMTs in order to ensure compliance with environmental protection obligations and incorporate findings into the development of more

---

[28] The term "Pilot Mining Test" is used in this section as an umbrella term to encompass all forms of testing referred to in UNCLOS, the Exploration Regulations, the Draft Exploitation Regulations and the Discussion Paper on the Draft EnvRegs. The term itself is not specifically used but was instead chosen because it is widely used within the field by different stakeholders but without an agreement of what it actually entails legally. Terms used include "testing of mining systems" (Annex III, Article 17 (2)(c)), "testing of recovery systems" (Exploration Regulations), "testing of equipment" (Annex III, Article 17 (2)(g)), "testing of plant" (Annex III, Article 17 (2)(g)), "testing of processing facilities" (Annex III, Article 17 (2)(c)), Exploration Regulations, "production tests" (Draft Exploitation Regulations), "capacity tests" (Draft Exploitation Regulations), "testing of transportation systems" (Exploration Regulations) and "testing of collecting systems" (Discussion Paper). Likewise the term "assessment of technological developments" (Implementing Agreement, Section 2 (1)) must also be considered in this context as well as broader legal definitions of "installations", "equipment", "devices" and "systems".

environmentally-friendly processes. More fundamentally, PMTs are a prime opportunity for better understanding the consequences of human impacts on the sensitive environment of the deep sea and enabling early intervention by contractors, sponsoring States, and the ISA to prevent serious harm to the marine environment.

A number of legal issues become apparent when attempting to regulate PMTs, however. The first concerns the nature and purpose of an obligation to conduct test mining - if one in fact exists. This is directly related to an emerging obligation to use best available techniques during operations. The second involves the lack of definitions of the "objective criteria" for determining the scale of equipment, systems, and testing processes, which is essential for determining when different regulatory steps must occur during pilot mining. This requires disaggregating the technological development process into multiple steps in order to regulate testing at the transition stage between exploration and exploitation and to better control the multiple types of tests making up a comprehensive mining test. Without clearly defining the criteria to be tested in each step, it is impossible to design a targeted, multi-phase EIA process to ensure compliance with environmental protection obligations in each stage of testing. Further, despite the "fundamental link" between exploration and exploitation, which underpins UNCLOS' approach to regulating deep seabed mining, there is a lack of mechanisms in the ISA's decision-making process to make meaningful use of the results obtained during test mining in order to determine whether commercial-scale mining should proceed at all. In particular, it has been largely ignored to date that the exploitation phase, in fact, involves two specific stages – development and commercial production – with distinct legal characteristics. This creates an opportunity to introduce an additional review and decision-making mechanism prior to commercial production, as opposed to the current mechanism, which authorises all exploitation activities including production, before the development phase has even begun.

Questions for discussion

- Is there a duty to conduct pilot mining? If so, which testing activities may be conducted in the exploration phase vs. the exploitation phase?

- How can pilot mining be used to determine Best Available Techniques (BAT)?

- Is an international standard for technical readiness levels (TRLs) a feasible method for regulating mining tests during exploration and exploitation?

- Can TRLs be used as a structure for introducing an increasingly stringent, multi-phase EIA requirement specifically to PMTs for contractors, sponsoring States and the ISA?

- How can a mechanism be incorporated into the Environmental Regulations and the Exploitation Regulations to ensure that the ISA receives sufficient information about PMTs to exercise effective control over the development of mining technologies and their environmental impacts?

- Can a production authorisation be introduced in the exploitation stage to enable the ISA to decide whether commercial production should begin?

Options to be deliberated

- Creation of a standard (or adaptation of an existing standard) to define TRLs in order to better regulate PMTs;

- Incorporation of a multi-phase EIA requirement in the Draft Exploitation Regulations and Draft EnvRegs in order to address critical environmental issues at each stage in technological development as early as possible in the development process and prior to commercial production;

- Inclusion of a production authorisation process after conclusion of the development stage of exploitation – reflecting current environmental protection requirements, including the precautionary approach – to enable the ISA to review the environmental impacts of PMTs prior to authorising any large-scale mining to proceed.

Recommendations, including potential next steps

- Agree on a standard for defining TRLs, including specific guidelines and criteria for contractors with respect to each phase;

- Agree on the structure and content of EIA obligations specifically applicable to PMTs and how these shall be taken into account in the overarching EIA process concerning commercial production;

- Incorporate a production authorisation process into the Draft Exploitation Regulations and Draft EnvRegs to ensure compliance with all environmental protection requirements and, if necessary, prohibit commercial production if serious harm to the marine environment has occurred during PMTs.

Interventions following the presentation

An intervention pointed to the fact that the term "pilot mining" was not used in the Convention or the ISA Regulations. It was emphasised that the current Exploration Regulations allow contractors to carry out tests of equipment or mining systems at any scale as desired, as long as the guidance to contractors issued by the LTC was observed. Beyond a resource-specific spatial impact on the seafloor, the provisions include the obligation to provide an EIA to the LTC. However, such testing is not mandatory. Another discussant expected that although there is no legal obligation to carry out mining tests during the exploration phase, this is very likely to happen because the demonstration of technical readiness was a precondition for raising capital; also, the inspectorate would later ask for an operations plan, including details and certification of all instruments.

Another intervention addressed the long list of environmental parameters to be measured prior to and after carrying out tests, as indicated in the LTC guidance for contractors.

The last intervention noted that it was yet to be decided how testing of any scale could be more explicitly included in the Exploitation Regulations. In this intervention, a need was seen to clarify whether a 2-step process as suggested in the presentation may be a useful approach to ensure that an effective environmental protection from the harmful effects of testing was demonstrated. In the intervention, it was suggested that the currently defined requirements for e.g. baseline surveys for the exploration context may need revision for their use in the context of exploitation applications.

7.5.2 Working Group Discussions

How can Pilot Mining Tests be integrated into the regulatory process / the regulations?

Based on the previous interventions and issues raised during the discussion (CHAPTER 7.5.1 above), the question was reformulated to: How can **equipment testing** be integrated into the regulatory process/the regulations? The issues discussed were:

- There was a view "that pilot mining and test mining have different connotations and different meanings depending on who is using it". However, it was not clear whether all of this subsumed under the heading of equipment testing, to use the terminology in UNCLOS. How do we assess the environmental and human impact if the focus is on equipment testing?

- It was noted that contractors are allowed to carry out tests during exploration. In light of this view, it was held that those tests will increase in size, i.e. can be part of an evolutionary process, and the implications for the corresponding EIAs are that these will grow in parallel, too. This view also held that an exploitation contract would be conditional to the results of such tests bearing in mind that exploration is very time consuming;

- The group was of the view that an EIA was a prerequisite to granting an exploitation contract but this was seen as an ongoing process from exploration transitioning into exploitation. In this sense, it was assumed that there would have been EIAs for field tests along the way during the conduct of exploration activities, so that by the time of the start of the exploitation phase, an EIA would be based on the earlier equipment tests;

- A question was raised, whether the function of testing may be to confirm/reject the credibility of the exploitation model being employed?

- Objectives need to be established between the ISA and the contractors, with an indication of how the contractors will achieve those objectives based on the contractor's rate of work, etc.;

- A view considered that there was a clear indication of the need to reduce regulatory uncertainty, and hence that it would be important for the contractor to obtain an exploitation contract in order to continue its work. A view was also held that that risks lie with the contractor;

- The LTC in granting exploitation contracts needs to consider a broader set of parameters, i.e. to ensure that the contractor has the financial and technical expertise to pursue exploitation. It was said that it was in the contractor's interest to ensure that the equipment meets international standards, because it is a commercial undertaking. It is not the task of the ISA to dictate the type of equipment being used – this should depend on the market availability;

- The main conclusion of the working group was that the details of the regulatory process in relation to equipment testing require clarification.

Can a Multi-Phase EIA Process be Installed for Test Mining?

It was considered that the combination of (1) the EIA(s) required for equipment testing during the exploration phase, and (2) the subsequent EIA prepared as part of the application process for exploitation, effectively represents a multi-phase EIA process. The view was held that this is sufficient to allow for proper functional testing of equipment and its potential impact on the environment and that the results of such tests would be most useful to integrate into the EIA of an exploitation application, so no additional steps would be required in this regard. The issues was raised that contractors, in fact, will be highly motivated to thoroughly address the environmental impacts of their technologies, as the performance, monitoring, and reporting on test(s) will form a key component of their application for exploitation and will have to convince the ISA that the applicant has gained the required knowledge on ecosystem impacts associated with its planned operations. A view noted that there is a limit to what you can ask from contractors, and that the industry may refrain from mining if requirements for additional, multi-phase test mining and EIAs are implemented.

Moving somewhat away from the prescribed question on multi-phase EIAs, the discussion turned to analysing the characteristics of mining equipment as foreseen by industry and how testing of these can inform EIAs. One view considered that contractors are starting to communicate with each other, which is an important first step to using BAT. In this sense, the focus of mining technology tests will be on collectors. It was also mentioned that no commercial contractor is planning to test riser systems before transitioning to commercial-scale mining as these are expensive and relatively predictable (appropriate riser technology had already been developed in the 1970s).

The discussion shifted to what scale of spatial and/or equipment testing would be necessary to obtain a realistic view of environmental impacts, i.e. what may be missed if tests fail to represent full commercial-scale mining or lack proper regulation and documentation. One view held that if initial equipment testing is carried out in shallower waters, additional test mining should take place in the relevant ecosystem (e.g. CCZ) at full depth. The scales of tests and mining operations (size of equipment, area impacted, duration) may affect ecosystem response – not necessarily in a linear way. The scale of test mining required to generate ecosystem impacts that are representative of commercial mining is not clear. Also the separate tests of components may result in impacts that do not necessarily reflect the impact of assembled operational systems.

In addition, a view was held that equipment tests provide a unique opportunity not only to check for equipment performance but also to test protocols and technologies for monitoring and how to make best use of the data (e.g. in terms of creating meaningful data products that help in assessing the impact). One opinion considered that repeated, and well-regulated test mining may facilitate the generation of knowledge in this field. The issue was also raised that properly monitored test mining will increase environmental knowledge that not only feeds into project-scale EIAs but also strategic assessments for the entire region. A view was held that appropriate protocols and technologies for monitoring can be tested so as to make best use of the data, and new insights can flow directly back into the regulatory process. Furthermore, the view was expressed that baseline requirements, as laid out in the regulations for exploration should be updated to represent current knowledge on relevant indicators. The best possible baselines are required as a benchmark to detect impacts and effects associated with test mining.

The discussion turned to how well the regulations cover things that go wrong. How can contingency plans be set up, when should they be implemented, what are the consequences? Even with the most thorough planning and testing, unforeseen complications may occur during mining operations, such as equipment failure or unexpected adverse effects on the environment and its ecosystem(s). There are other issues connected to this, e.g. concerning triggers, responsibilities, and actions to take, that need to be sorted out and that should be included in regulations and management plans.

The discussion ended on the note that it would be helpful to have PRZs and IRZs in place in order to assess the environmental impacts of test mining during the exploration phase.

Report of an Additional Working Group on Pilot Mining Tests

During this additional working group on PMTs, participants shared the more technical perspective, comparing countries' experiences with developing new technologies. Several points emerged:

- There are some unresolved terminology issues: "equipment tests" (UNCLOS) are not necessarily the same thing as "production tests" or "capacity tests" (Draft Exploitation Regulations);

- There is an urgent need to define what BAT is and how to achieve it - as opposed to "optimal technology", which would set completely different requirements;

- Whereas in the working group on multi-phase EIA processes the opinion was raised that risers would work and the functions only need to be modelled (see sub-chapter above), the practical experience of some of the participants involved in technical development here was very different;

- It was felt that many lessons could be learned from the oil and gas industry, in particularly from their cooperation experience (benefits of shared risk, shared success);

- Monitoring: BAT should make use of advanced technology (ROVs) that is already available, e.g. precision sampling;

- Expand on impact and restoration experiments – as results often only become visible on decadal scales;

- It should be considered how confidentiality clauses concerning proprietary information affect access to environmental information (special issue for PMT);

- It should be considered how modelling can best support evaluation: continual improvement of monitoring, mitigation measures, emergency responses;

- What are the criteria for a successful PMT?

- Future steps depend on how testing is defined (PMT, equipment test, testing, etc.).

Report of an additional Working Group on incentives for contractors

The working group discussed options for incentives for innovations and adoption of practices and technologies which are "above and beyond" the contract obligations, meaning that such measures would not be legally required.

Three groups of potential sources of incentives (preferences or rewards systems) were differentiated:

1.) The ISA (internal), through *inter alia*

- Regulatory streamlining, such as relief in reporting requirements (frequency) or reduction in bonds;

- Reduction of royalties rates (issues with balance of CHM: return on minerals vs. lesser degradation of the environment);

- Priority consideration for a future partnership with the enterprise (access to reserved areas);

- Access to closed-off areas (e.g. through use of low-impact technology);

- "Good" contractor list.

2.) External, through *inter alia*:

- Green commodities preference – distinction based on innovations, technologies and practices. Challenges would be that these are expensive to create, assume a critical market share of deep-sea minerals, need third-party certifications;

- Carbon offset scheme;

- Role of the sponsoring State: National treatment preference for "good contractors", tax releases, reduced insurance fees;

- Preferable treatment by banks/lenders.

3.) Inter-contractor, through *inter alia*:

- Exchange of experiences and knowledge, benefits from sharing.

7.6 Regional Governance

7.6.1 Presentation: Overarching Issues Around Regional Governance of Deep Seabed Mining
Daniel Jones and Phil Weaver

It is recognised that the ISA will need to develop its approach to regional governance to ensure that the impacts of mining are seen in the context of the global ocean. Following best practice in other offshore industries, such an approach is likely to involve a tiered structure, with an over-arching strategy that sets the requirements and processes by which Regional Environmental Assessment (REA) should be carried out. The strategy can be developed using Strategic Environmental Assessment (SEA). SEA is a high-level planning process, developed to evaluate the long-term environmental impacts and effects of multiple sectors and activities[29]. Ultimately, SEA should help protect the environment and promote sustainability by integrating environmental issues in decision making. SEA is particularly appropriate where there will be multiple contracts in one region and/or where there are multiple activities taking place at the same time.

---

[29] Therivel, R., 2010. Strategic Environmental Assessment in Action. Earthscan, London.

Regional Environmental Assessment (REA) involves the collection of information on a particular region and an assessment of that region from which regional environmental management plans can be produced. An assessment is the formal decision-making process during which the optimal strategic approach is determined to limit the effects of activities based on all available data and information. This decision-making process is usually documented in an SEA or REA report. The assessment leads to the formulation of a management plan (SEMP or REMP) to provide a practical roadmap to deal with those effects so as to limit their impact. For the ISA, REA could be conducted at ocean basin scale or smaller if the ecosystems or habitats are complex and frequently changing. The REA then sets criteria and objectives for the local scale (or contractor block) activities that are part of the Environmental Impact Assessment (EIA). The tiered approach is shown in Figure 7.

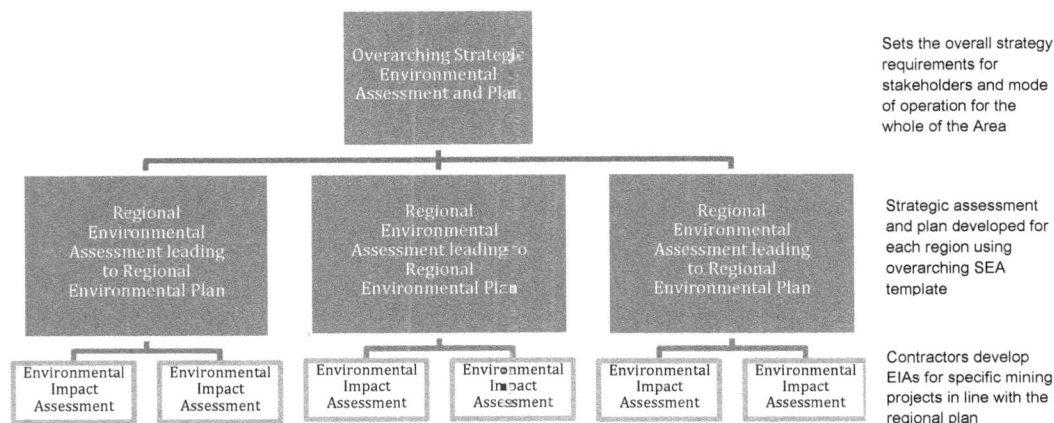

Figure 7: *Model for environmental planning and management at different levels within the ISA, showing the relationships between the different tiers of assessment and planning.*

The REMP will best be compiled under the guidance of an overarching SEA and its development should involve the ISA, contractors, scientists, and regional organisations plus any other stakeholders. Ideally a REMP should be in place before the award of exploration contracts, but data from contractors may be critical to its development, in which case it should be in place before the award of exploitation contracts. However, the ISA could in addition call for national funding agencies of member States to fund and equip regional environmental baseline investigations, either collectively or individually. The REA and REMP should set clear goals and objectives for the local scale activities that are controlled by the agreement between the ISA and the contractor. These will then form part of the contractor's EIA and Environmental Impact Statement (EIS). Monitoring of environmental status will be required (including at sea) to assess the effectiveness of the measures as set out in the combined SEMP, REMP, and EIS. As part of the management cycle, this may lead to periodic modifications of the plans and statements to optimise the environmental status, as aimed at in the environmental objectives.

As a minimum, REMPs should include the following elements, many based on the content of the high-level SEA and more specific REA:

- Gathering of baseline data from multiple sources;

- Establishment of a network of APEIs based on sound scientific design principles such as those used in defining the CCZ EMP[30]. The APEIs should be protected from mining activities in perpetuity;

- Spatial planning, taking into account the genetic connectivity of populations, biogeographic zones and finer-scale gradients including seasonal and inter-annual variability;

- Establishment of smaller-scale protected areas within contractor areas, based on both representativity criteria and the occurrence of Ecologically or Biologically Significant Marine Areas and likely hot spots of endemism or biological significance (e.g. as suggested by geomorphology, oceanography, biological or geological processes);

- Vulnerable Marine Ecosystems and any other areas where there are prior conservation measures in place should be identified and made provision for;

- Identification of cumulative and synergistic impacts (e.g. multiple mining operations, ocean acidification, global warming, fisheries);

- The ISA should take account of other relevant uses of the ocean while exercising its own competences (Article 147 paragraphs 1 and 3 of the Convention). Address environmental limits and thresholds;

- Mitigation measures may primarily be through spatial planning approaches, but are not limited to these (potentially including activity management or temporal management approaches);

- Mechanisms need to be built into the REMP to ensure independent scientific research and monitoring to ensure that the objectives of the REMP are being achieved. Results will need to feed back into refinements of the REMP or other aspects of adaptive management.

In order to progress the work towards REA and REMPs, the following issues have to be resolved:

- How can the SEA process be integrated into the practice and policy of the ISA?

- What should be the timeframe for strategic initiatives (SEA and multiple REAs) to be developed?

- How will the SEA process best link with EIA and claim-scale activities?

- How can the evidence base that underpins SEA be collected and openly shared?

- What is the minimum amount of data required to perform an REA?

- Should the focus be solely on spatial environmental management approaches?

- How can SEMP and REMP be made legally binding, particularly after contracts are issued?

- Who will conduct SEA / REA and how will they interact with the ISA and other stakeholders?

---

[30]   Wedding, L.M., Friedlander, A.M., Kittinger, J.N., Watling, L., Gaines, S.D., Bennett, M., Hardy, S.M., Smith, C.R., 2013. From principles to practice: a spatial approach to systematic conservation planning in the deep sea. Proceedings of the Royal Society B: Biological Sciences 280 (1773), 20131684.

- How should the development of SEA / REAs be funded? Who should fund baseline data collection and who should fund monitoring?

- What happens if SEA identifies management strategies that affect mining claims, e.g. limits being put on the total amount of mining in a region?

- Which are the priority areas for establishing REAs and what scale is most appropriate (e.g. do we need more than one for the CCZ)?

- Should specifically tailored SEA / REA guidance protocols be developed for the ISA?

- What mechanisms should be established for revision of REMPs and their relationship to adaptive management? What should be the frequency of SEA / REA review and how will the effectiveness of the SEA / REA measures be determined?

- How can the effects of cumulative impact be taken into account?

7.6.2 Presentation: Regional Governance and the CCZ Environmental Management Plan
Phil Weaver and Daniel Jones

The Environmental Management Plan (EMP) for the Clarion–Clipperton Fracture Zone (CCZ) in the Eastern Central Pacific was approved for an initial three-year period by the ISA in 2012 (ISBA/18/C/22)[31]. It applies to an extensive region of the Pacific, rich in polymetallic nodules, located beyond national jurisdiction at water depths of 4,000-6,000 m (ISBA/17/LTC/7)[32]. The CCZ-EMP is the first regional-scale environmental management plan for the deep seabed and exempts roughly 25% of the CCZ management area from exploration licensing for certain periods. However, the boundaries of the area could be defined quite differently depending on whether they are drawn around the main area of contractor blocks, the bounding fracture zones, or to include the conservation areas known as Areas of Particular Environmental Interest (APEIs). The CCZ-EMP provides a proactive spatial management strategy that is expert-driven and that seeks to conserve a representative fraction of the region from anticipated impacts associated with the mining of polymetallic nodules by designating a series of nine APEIs. Each APEI has a core area 200 x 200 km surrounded by a 100 km buffer making each APEI 400 x 400 km in size (40,000 km^2).

The CCZ-EMP does not foresee any further spatial or non-spatial measures and does not relate to a defined environmental baseline. However, a plan has been agreed to provide a quality status report of the region every 5 years. The guiding principles of the CCZ-EMP are the common heritage of mankind; the precautionary approach; protection and preservation of the marine environment; prior EIA; conservation and sustainable use

---

[31] International Seabed Authority, 2012. Decision of the Council relating to an environmental management plan for the Clarion-Clipperton Zone ISBA/18/C/22. International Seabed Authority, Kingston, Jamaica.

[32] International Seabed Authority, 2011. Environmental Management Plan for the Clarion Clipperton Zone. ISBA/17/LTC/7. International Seabed Authority, Kingston, Jamaica.

of biodiversity; and transparency. The design principles underpinning the CCZ-EMP are described in Wedding et al. (2013)[33].

Currently, the CCZ-EMP only relates to the APEIs, which lie outside of the contractor claim areas. A more comprehensive approach would include conservation measures throughout the CCZ region, including in the central area between the two fracture zones where the contractor claims are concentrated. Preservation Reference Zones, which are not impacted by mining, will need to be established in each contractor claim and these should form part of the CCZ-EMP, including more detailed guidance on how these should be established and managed.

Data are critical to the implementation of the CCZ-EMP. In particular, understanding possible endemism and extinction of localised species are important considerations. To this end a partnership approach requires:

- The scientific community: to determine parameters for ecosystem connectivity; to evaluate re-colonisation potential; and to incorporate where possible additional data sources; and

- Contractors: to collate physical and biological data in order to achieve biogeographical baselines (however, it should be noted that as yet contractors are under no legal obligations to supply baseline data for APEIs).

The CCZ regional environmental management plan was reviewed by the ISA in 2016 (ISA, 2016)[34]. The review identified a number of actions for both the ISA Secretariat and the contractors and listed 13 measures included in the CCZ-EMP, of which only two - the creation of the areas of particular environmental interest and the convening of the three workshops on taxonomy, have been implemented.

The main issues to be addressed in the CCZ-EMP can be summarised as:

- The paucity of data for the whole CCZ on which the plan was based. This can only be improved by the open sharing of contractor-collected environmental data; however, the contractors are under no obligation to collect data outside of their contract areas, including in the APEIs. Recently contractors have collected data in five of the APEIs and scientific cruises have provided additional data in two APEIs. However, no data at all has been collected in three of the APEIs;

- The duration of the APEIs, which need to be in place until mining has ceased in the CCZ, owing to the very slow recovery rates of the ecosystem, but are already up for review even before mining begins;

- Mechanisms for monitoring e.g. to ensure that there is no impact from mining on the APEI core areas and to assess how effective the APEIs are in contributing to the environmental goals under which they were established. There is no agreement as to who will pay for subsequent monitoring in

---

[33]   Wedding, L.M., Friedlander, A.M., Kittinger, J.N., Watling, L., Gaines, S.D., Bennett, M., Hardy, S.M., Smith, C.R., 2013. From principles to practice: a spatial approach to systematic conservation planning in the deep sea. Proceedings of the Royal Society B: Biological Sciences 280 (1773), 20131684.

[34]   International Seabed Authority, 2016. Review of the implementation of the environmental management plan for the Clarion-Clipperton Fracture Zone. ISBA/22/LTC/12. International Seabed Authority, Legal and Technical Commission, Kingston, Jamaica, pp. 1-10.

the APEIs or who will carry it out. Nor has the frequency of required monitoring been established or who will assess the results therefrom;

- The effectiveness of the APEIs as part of the overall conservation measures for the CCZ, which will include more local activities as part of the EIA process. These could include networks of Preservation Reference Areas (PRZs) in the contractor areas, though these would need to be of sufficient number and large enough to achieve conservation goals, and be in place as long as the APEIs;

- How the CCZ EMP fits within an overarching SEA that is yet to be developed by the ISA;

- The possible need for sub-regional EMPs in such a vast area as the CCZ. The CCZ covers an area of over 5 million square kilometres, encompassing several different ecosystems;

- Mechanisms for working with other relevant international organisations and eventually coastal States in waters adjacent to the Area to establish agreed common principles for conservation measures;

- The need for the process to be open and transparent.

7.6.3 Plenary Discussion on Overarching Issues Around Regional Governance of Deep Seabed Mining and the CCZ-EMP

The first comment following the two presentations summarised in CHAPTER 7.6.1 and CHAPTER 7.6.2 highlighted that regional environmental assessment and the development of regional management plans will require the coordination and cooperation with other mechanisms and processes in the region, for example in the Indian Ocean. It was noted, also, that sponsoring States have signed up to a multitude of bilateral and multilateral international and regional agreements and commitments. The question was how this could be coordinated. In this regard, the views focused on the ISA being the owner and driver of the process (at least from a scientific perspective), although it was assumed that this will take a long time to establish.

Regarding the proposed scale on which a SEA should be undertaken, a view was held that an overarching SEA was required on a global scale, which could focus on technologies, processes, and interactions between actors and different levels of the overall governance approach.

Several questions were related to the envisaged sequence of procedural steps:

- The concern was raised that only contractors would be able to provide data on the scale required for the management of regions, and that this would counter the proposed top-down sequence of planning steps in the tiered approach. Whereas ideally, contractors should be able to deliver sufficient background data on their contract areas, they do not have obligations with respect to APEIs. An opinion held that modern scientific instruments and methods will allow for large-scale, long-term, and continuous observation in remote areas, e.g. as developed in the Deep Ocean Observation System[35]. A view was expressed to the effect that climate change would be taken into account as shifting baseline and as one of the cumulative impacts in the REMP.

---

[35] http://www.deepoceanobserving.org/

- In the CCZ, the designation of APEIs occurred after contracts for nodule exploration had been awarded - should it be possible to develop the regulations in such a way that a requirement that SEAs, REAs, and REMPs be in place at least prior to the first exploitation contract is installed? This was found to be desirable as ideally, the REA should suggest processes and ways that need to be addressed in project-specific EIAs (the EIA should be produced in the light of the REA).

- The view was expressed that the question remains as to what process can deliver an integrated regional management plan which takes account of all interests and provides a representative network of protected areas (MPAs, APEIs). A top-down approach was recommended for the design of the management system to avoid unnecessary revisions.

An opinion was raised indicating that lessons can be learned from examples for implementing SEAs and REAs in other industries and legal contexts: e.g. the UK aggregates industry has made a voluntary commitment to undertake REAs[36] and subsequently develop monitoring programmes for a number of strategic areas of offshore extraction.

7.6.4   World Café Discussions

Is a multi-tiers approach to management required?

The view was expressed that a global, higher-level SEA/SEMP is required and that it can be extremely useful, although there was also the view which considered that an SEA was not good use of limited resources, as much of the information in it already exists elsewhere in available policy documents. It was felt that SEA/SEMP substantiates the best practice available, and can provide the framework for regional and local-scale planning as well as delivering coherence and integrity to marine ecosystems. On the tier level below (REA/REMP), the participants felt that this is a highly valuable and important exercise to do.

Benefits of the multiple-tiers approach to management are, among others:

- Marine ecosystems need a hierarchial approach to management;

- Higher-level management sets the framework and consistent objectives for action at regional and local levels: The SEA/SEMP sets global, overarching goals, methods and techniques; REA/REMPs will be mineral-specific and contain adequate thresholds; project-specific EIAs/EMMPs with corresponding mitigation strategies are informed by the previous tiers (contractor level);

- Regions, stakeholders, and relevant cumulative impacts are defined and addressed, alternatives are considered, and objectives, thresholds, processes, and choices are laid out clearly;

- A multi-tiers approach will increase certainty for all actors;

- Tiers represent an information hierarchy from simple/general to complex/greater specificity;

---

[36] see e.g. http://jncc.defra.gov.uk/pdf/rea%20framework%20guidelines_final.pdf . In 2002 the UK Government imposed a levy on all primary aggregate production (including marine aggregates) to reflect the environmental cost of winning these materials. A proportion of the revenue generated is used to provide a source of funding for research aimed at minimising the effects of aggregate production. This fund, delivered through Defra, is known as the Aggregate Levy Sustainability Fund (ALSF). See further http://jncc.defra.gov.uk/page-4279

- REAs provide overarching regional information that is valuable for the project-specific EIA process (reduces risk and may help contractors in selecting their mining and reference sites). This might also be an incentive for contractors to monitor APEIs alongside their license areas.

However, a call was made to keep it simple. UNCLOS sets high-level objectives, which could greatly help assessment processes, but at the same time, a view was held that UNCLOS does not provide for the development of a SEMP, making its legal base questionable.

It was stressed that linkages between tools (SEA/SEMP, REA/REMP, EIA/EMMP) is fundamental. Some participants suggested the need for an intermediate tier between the SEMP (all minerals) and the REMP (one particular mineral), taking the ocean basin as relevant spatial scale (e.g. REMP for all minerals in the Pacific; REMP for all minerals in the Atlantic, etc.). Furthermore, it was argued that strategic and regional plans should not be set in stone, but should be reviewable and responsive to new information as it becomes available (iterative process true to the precautionary approach). Feedback loops across the tiers were considered particularly important (top-down as well as bottom-up). However, there is a requirement of minimum data for any of the tiers and both sources of data and funding were stressed as important obstacles. An opinion considered that existing information was considered to be both scattered and incomplete. As potential sources of data, there was a suggestion to engage the global observing community (e.g. GOOS; Deep Ocean Observing Strategy; ARGO buoys). Sponsoring States can perform an important role in data collection as well, for instance, by making use of their cargo ships, etc. There are numerous examples of SEA and REA processes in the national context that could be used as blue print.

Several questions were raised. How can a classic model of SEA be adapted to a new activity, such as DSM? How can activities other than mining and socio-economic issues be included in the SEA/SEMP? What are the limits of the ISA mandate? How binding and/or adaptive should regional plans be - do contractors have to follow them? What if national standards/agreements are different? Can sponsoring States bring in other regional agreements to which they are parties? What happens if there is a conflict between objectives at different scales? How are inconsistencies handled? What happens when States have not signed all international agreements (e.g. Aarhus, flag State agreements, IMO, etc.)? A suggestion was made to introduce a clause into the Exploitation Regulations requiring compliance with REMPs, and mandatory adaptation of mining plans when the REMPs are revised.

What are the pros and cons of regional governance?

The commonality of the comments made by the participants was on the pros and cons of regional environmental management rather than on regional governance. In principle, no one spoke against the idea of regional management. While many pros were raised, some cons were also mentioned.

The participants took due account of the Discussion Paper on the Draft EnvRegs, in particular Section 11 concerning strategic environmental planning and management (p.16). Consequently, they used the CCZ-EMP as the starting point of the discussions. They also looked at the question as to whether issues related to regional management should be included in the context of the Draft EnvRegs, or follow the current approach for the CCZ-EMP, i.e. a policy document developed by the LTC and endorsed by the Council. On this issue, a view considered that these aspects should be included in the Regulations, while another view favoured a flexible and ad-hoc approach in light of the current EMP in the CCZ.

Pros

- Regional management is good and essential for environmental protection, and could prevent environmental transboundary impacts;

- It promotes ecosystem-based management (creation of a network of APEIs) and constitutes best environmental practice (BEP), which enhances the confidence of the contractors. It was also mentioned that it can help standardisation, capacity-building, and stakeholder participation;

- It promotes uniform and consistent approaches across regions. However, one observation was made that regional management takes into account the heterogeneity of the marine environment;

- It may also assist in designing buffer and core areas within the PRZ;

- It encourages cross-sectoral cooperation (RFMOs, regional frameworks) and assists in implementing the reasonable regard and due regard provisions found in UNCLOS (Articles 87(2) and 147).

Cons

- It was mentioned that the process could be lengthy, cumbersome, and costly. It could also give rise to additional obligations to contractor and sponsoring States;

- Another con was the current legal uncertainty on the status of environmental management plans;

- The ISA has limited capacity at the moment to further develop regional environmental management plans;

- There were remarks in the sense that regional management in the Area is different from experiences at the national level.

Suggestions made by participants

- Begin incrementally through a gradual approach;

- On the issue of costs, it was mentioned that incentives could be considered for contractors so they could research and collect data outside their contractor blocks in order to enhance/develop regional environmental management schemes;

- It was suggested that, in regional management, all States Parties to UNCLOS, not only contractors, should take ownership.

What are objective criteria for initialising regional governance? Who is in charge?

The aim of regional governance is to ensure coherence and application of globally agreed "good governance principles" at regional scale in regional governance processes, ensure environmental protection in line with conservation objectives and use, including the conservation of mineral habitats, and awareness-raising for areas of environmental significance, etc. Objective criteria defining good regional ocean governance include, amongst others, the implementation of the ecosystem approach, the precautionary principle, and the polluter-pays principle and the subdivision of a region in line with the configurations of the ecosystems, including the creation of conservation objectives as part of the ocean governance structure. A regional-scale approach to governance and management was seen to be essential because multiple actors may cause impacts beyond their immediate mining area, affecting areas in other jurisdictions, other contract areas, or sensitive/protected areas and potentially causing cumulative impacts.

Who is responsible for regional governance, and how?

There was the view that, with respect to DSM, this regional governance was the responsibility of the ISA, shared with contractors and sponsoring States. It was also believed that non-sponsoring States should engage more, based on their duty to promote and support marine scientific research, and as they will share in the "benefits for mankind" from mining. The ISA could expand its capacity to better include input from scientists and stakeholders by making greater use of States Parties and their independent scientists, by including scientists in delegations for the Council, and/or initiating working groups of State representatives and/or environmental experts who work on specific issues throughout the year.

It was acknowledged that the ISA does not have the financial capability to undertake all the necessary research needed for good regional governance. The role of ISA can be that of a facilitator, identifying the different actors and their contributions to regional environmental assessment and management plans. The ISA was seen to have an obligation to promote marine scientific research and thus encourage collaborations with sponsoring States and contractors to provide scientific input to e.g. APEIs and REMPs, i.e. also for areas beyond their immediate commercial license areas, in order to better understand the wider array of regional processes and functions. This, in turn, will help contractors to meet their own obligations for project-specific EIAs and EMMPs.

There should also be wider cooperation with other sectors and organisations, including citizens and scientists. It was suggested that the ISA's lack of financial capabilities may be addressed by MoUs with related international and regional organisations, potentially opening up for new sources of funding. Furthermore, a need for enhanced, independent science cooperation was identified. Last but not least, it was mentioned that there is potentially room, under the BBNJ Implementing Agreement currently under discussion at the UN, for creating a larger-scale process for strategic environmental management and planning at ocean basin scale.

REAs and REMPs are crucially dependent on scientific data. Input of environmental data from contractors, other organisations, and from the scientific community will require centralised collecting efforts as soon as possible and analysis will be a long-term iterative process. To make regions operational, it was considered important to match the areas of interest in the mineral resource with the respective biogeographic subdivision – as these may not necessarily be the same.

When should regional governance be initialised?

There was the view that regional ocean governance processes are required to set the context for future exploration and exploitation claims. The timing, however, was subject to debate. Ideally, the latest point for triggering the process would be when the first prospecting notification for a particular area is received by the ISA, in line with UNCLOS Annex III, Art. 2. This would initiate a process of data collection, encourage marine scientific research, and require the prospector or exploration contractor to start gathering (missing) data in and beyond its contract area. Another trigger could be the receipt of more than one application for exploration, i.e. the urgency of regional management processes should intensify as numbers of prospectors/exploration contractors intensify. Another view was that governance processes are only required when the density of exploration license areas is high in a part of the Area. The participants felt that the process should begin now, under the ISA's mandate to promote marine scientific research and marine environmental protection, conservation, and management. At the latest, it should be in place before exploitation contracts are authorised.

7.6.5   Presentation: Spatial Management Approaches
Daniel Jones and Phil Weaver

Spatial management in the oceans includes both marine spatial planning and the ongoing management of marine areas. It is a practical way to create and establish a more rational organisation of the use of marine space and the interactions between its uses, to balance demands for development with the need to protect marine ecosystems, and to achieve social and economic objectives in an open and planned way[37]. Good spatial management is ecosystem-based, integrated, area-based, adaptive, strategic, and participatory. Spatial management has many benefits including environmental, economic and social benefits[38]. Spatial management is not the only approach available, management can for example also be based on limiting specific activities or technologies, or preventing impacts at certain times. There are several mechanisms for introducing spatial management initiatives, including policy, strategic planning (linked to regional environmental management plans), project planning (linked to project environmental management plans) or voluntary initiatives as part of company policy. One of the great benefits of spatial management is that protecting space is simpler than setting other management approaches. A wide range of biological and physical parameters increase with area, for example population sizes, biodiversity, habitat heterogeneity, and by protecting areas, particularly large areas, it is likely that some of this variation will be captured. Clearly, with more knowledge on the ecosystems, targeted spatial management approaches can be more effective.

A major review of marine protected areas has suggested that five factors are important to defining the success of spatial management approaches. Effective protected areas are not impacted, well enforced, old, large, and ecologically isolated from other activities[39]. Approaches like this lead to evidence-based design criteria for spatial management approaches. The Convention on Biological Diversity suggests design criteria[40], emphasising protection of both special (ecologically or biologically significant) and representative areas, as well as highlighting the need for adequacy of protection (size of protected areas), replication of areas and connectivity / spacing of areas.

---

[37]   DEFRA, 2009. Our Seas-A Shared Resource-High-level marine objectives. Department of Environment, Food and Rural Affairs, London, UK, p. 12.

[38]   UNESCO, IOC, 2009. Marine spatial planning – A Step-by-Step Approach toward Ecosystem-based Management.

[39]   Edgar, G.J., Stuart-Smith, R.D., Willis, T.J., Kininmonth, S., Baker, S.C., Banks, S., Barrett, N.S., Becerro, M.A., Bernard, A.T.F., Berkhout, J., Buxton, C.D., Campbell, S.J., Cooper, A.T., Davey, M., Edgar, S.C., Forsterra, G., Galvan, D.E., Irigoyen, A.J., Kushner, D.J., Moura, R., Parnell, P.E., Shears, N.T., Soler, G., Strain, E.M.A., Thomson, R.J., 2014. Global conservation outcomes depend on marine protected areas with five key features. Nature 506 (7487), 216-220.

[40]   Secretariat of the Convention on Biological Diversity, 2004. Biodiversity issues for consideration in the planning, establishment and management of protected area sites and networks. CBD Technical Series no. 15. SCBD, Montreal, p. 164.

The underlying concepts for spatial management zones are similar for all types of mining. However, there are differences in considerations concerning the scale, spatial constraints, and ecology of these areas. The biological communities associated with active SMS deposits, for example, are very different from those in nodule fields, with the former being isolated areas with relatively high densities of fauna but relatively low diversities, whilst the latter are the opposite. Crusts are associated with typically diverse communities particularly of sessile suspension feeders and unlike the other two DSM resources may also be associated with commercial fish species. As a result of these ecological differences, design of management areas that enable monitoring and protection of the environment will necessarily lead to differences in size and location. However, many of the underlying design criteria will be similar across all deposits.

Spatial management approaches are currently envisaged at multiple spatial scales for deep seabed mining activities. At the regional scale a network of large marine protected areas, or Areas of Particular Environmental Interest (APEI), is anticipated, as has already been implemented in the CCZ[41]. This network of APEIs protects the greatest area of the seafloor (a total of 1,440,000 km^2) but is somewhat peripheral to mining activity, being arranged around the outside of the mine claim areas, the separation between individual APEIs is large (the shortest straight-line distances between APEIs range from 174 to 3017 km) and north-to-south separation of APEIs at a similar longitude is at least 407 km. As a result, these large areas may not be representative of mined areas and have limited efficacy in ameliorating impacts in the mined areas. Consideration needs to be given to the duration of APEIs, long-term protection needs to be balanced against adaptive management for this.

Within contractors' claim areas, spatial management may be carried out using the network of Preservation Reference Zones (PRZs) specified by the contractor as part of their environmental management plan. The PRZs are specified in the mining code as being "areas in which no mining shall occur to ensure representative and stable biota of the seabed in order to assess any changes in the flora and fauna of the marine environment". As such, they are monitoring areas. However, as they are no-mining areas they will need to be factored into spatial management and may have a role as conservation areas. The CCZ-EMP (ISBA/17/LTC/WP.1) indicates that the PRZ may have a "preservation" function. In both their preservation and monitoring roles, PRZs are intended to be representative of mined habitats and protected from the primary and secondary effects of mining activities. Further guidance will be developed to guide the placement, size, and nature of PRZs. We would suggest that consideration is given to having PRZs and IRZs targeted to direct impacts of mining as well as plume impacts. Furthermore, whilst PRZs should be representative of mined areas, other typical or vulnerable habitats in the claim area may need spatial protection.

It is also possible for spatial management measures to be included that are finer in scale than the PRZs. These could be implemented by changing the pattern of mining activities. For example, unmined corridors or patches could be maintained to mitigate impacts. The success of such approaches greatly depends on the nature and extent of secondary impacts, particularly plumes.

---

[41]   Wedding, L.M., Friedlander, A.M., Kittinger, J.N., Watling, L., Gaines, S.D., Bennett, M., Hardy, S.M., Smith, C.R., 2013. From principles to practice: a spatial approach to systematic conservation planning in the deep sea. Proceedings of the Royal Society B: Biological Sciences 280 (1773), 20131684.

7.6.6   World Café Discussions

Identification of required spatial measures to minimise environmental impacts - in addition to the designation of APEIs?

In order to identify possible spatial measures, geophysical structures/habitats should be categorised and mapped, as well as their associated biological communities and keystone species. The entire area that could potentially be impacted should be mapped by the contractor, including areas outside their claim. A network of protected areas should be established which should fulfil the following criteria:

- Long-term protection;

- Large-scale and appropriate for conservation objectives;

- Cover representative habitats and communities;

- Ensure connectivity.

It must be clarified whether PRZs serve conservation purposes or only reference purposes, and whether they can be mined at a later date. If PRZs are only for reference, new conservation areas within or between contractor blocks are necessary. These could be called "test preservation areas", should include areas with high densities of nodules and should not be impacted by plumes.

A systematic approach to spatial planning making use of all available instruments (APEIs, PRZs, IRZs) was proposed, which would support the development of REMPs for all areas to be adopted by the ISA. Further institutional mechanisms, including an environmental committee, may be necessary. In order to transform protected areas into effective MPAs and address cumulative spatial impacts, cooperation with other sectoral organisations should be initiated. This could involve links with VME activities and the BBNJ process.

Smaller-scale spatial measures were also suggested:

- Mining strategies for individual projects;

- Temporary closures or collection reductions, track pattern management to allow strips and corridors for species migration and recolonisation;

- Collection only from areas with high nodule density;

- Restrict activities to only part of the claim where nodule density is highest and leave representative areas in other parts of the claim unaffected;

- Spatial management of plumes to ensure that plumes extend over previously mined areas;

- Assess cumulative impacts together with other activities such as fishing;

- Vertical spatial planning in the water column to address waste and sediment disposal.

Further points included that no mining should occur on active vent sites, that climate change and ocean acidification should be taken into account, that it should be ensured that plumes do not affect PRZs and conservation areas in other claims, and that limits for rates of sediment re-deposition within licensed areas should be set.

Complement the required elements of the CCZ-EMP to derive a REMP template

Participants of the world café groups decided to focus on the good aspects, bad aspects, and the gaps in the CCZ-EMP. The points made below are in no particular order.

The review process as also foreseen for the CCZ-EMP was deemed very important. In the long term, management would need to balance the objective of stability on the one hand, e.g. for long-term conservation areas, and adaptability on the other hand, e.g. as a response to environmental change or changes in use patterns.

One important element was the setting of a timeline for the establishment of other REMPs. Although REMPs should ideally be in place as early as possible, it will have to be considered how to deal with potential conflicts arising when REMPs are adopted after exploration or exploitation contracts have been concluded. Concerns were raised on processes for biogeographic analysis and conflicts with other existing schemes.

A REMP should provide for the assumptions that environmental management strategies are based on, and the need to test these assumptions. Like the CCZ-EMP, they should have a clear objective / rule-based approach. However, these objectives will be different for each region depending on resource types, scales, activities, and data availability. Objective(s) around the size of APEIs have implications for efficacy, spacing, location, representativity, and conflicts and may need to be revisited also for the CCZ.

Who will administer REMPs in day-to-day management, data collection, and review? A much more specific identification of roles and responsibilities, including objectives for contractors and State Parties, should be detailed in REMPs. In practical terms, a REMP depends on an iterative process of data collection and submission, and so their availability should be sought.

REMPs should also set the agenda and process for defining PRZs/IRZs. This accentuates the need to consider relationships between zones, for example between individual APEIs (connectivity, representativity) and their relationship to PRZs/IRZs (recognising that PRZ/IRZs will only be defined at the end of the exploration or in the exploitation phase, and not all at once).

It is presently unclear how cumulative effects may be assessed. This relates to the question, how do the regional plans fit with other sectors (e.g. fishing), other impacts (e.g. climate change), and other jurisdictions (water column and national waters)?. The spatial management strategies adopted by the nearest coastal states should be taken into account, through consultation/cooperation/ coordination with relevant coastal States and including considerations of transboundary effects: between mine sites and between DSM and other ocean users.

Furthermore, REMPs should include:

- Obligations on contractors to support the achievement of the goals of the REMP, e.g. by providing data;

- Data obligations: enforcement of the submission of environmental Information, e.g. in annual reports – put obligation in the REMP, and a reference in regulations to make this aspect compulsory;

- Guidance to relevant elements requested for the EIA;

- Provisions that allow for claim modification (by negotiation) to allow designation of new or smaller APEIs (to achieve protection of representative habitats);

- Provisions for the need to collect data inside APEIs;

- It was noticed that the water column is inadequately dealt with in the CCZ-EMP, and it was considered important to integrate the effects of return water plume effects into REMPs.

Selection of areas as PRZ and IRZ – How, how many and where?

Prerequisites for the selection of PRZs and IRZs were considered to be: 1) good environmental baseline data collection, and 2) standardisation of the sampling approach in the PRZ, IRZ, etc.

How to define PRZs / where (several options):

- Grid-like multiple PRZs around the IRZ to monitor impacts;

- Network of PRZs depending on ecological conditions: requires good spatial representation of samples / enormous sampling effort and makes use of statistical analyses (e.g. MSD-plotting; Principle Components Analysis; ecosystem-based modelling) to determine the best positions of PRZs;

- Network of PRZs depending on spatial physical and/or habitat information (e.g. from multibeam, AUV imaging) - less time-consuming, better spatial resolution;

- PRZs should be located away from any impact area and preferably not near the boundary with neighbours, and the biota should be comparable to the one of the impacted area;

- PRZ location should ensure connectivity;

- PRZs need to contain high biodiversity, should be away from the plume to be intact by the end of mining, and should contain a similar exposed nodule area (for epifauna and epibionts);

- PRZs should serve as source populations, and they need to produce/provide at least as many juveniles as lost in (impacted) sink areas;

- According to one view, the PRZs should be different from APEIs, whereas others wanted to use similar criteria for the identification of PRZs as for the selection of APEIs;

- PRZs might need a surrounding buffer (as in APEIs) to isolate them from impacts (also from neighbours);

- The regulator might advise the contractor about neighbouring spatial planning plans to avoid transboundary impacts;

- A view was expressed considering that the identification should not be too prescriptive during the first stage and should allow for some flexibility, and that the regulator should take a regional perspective. There may be a need for additional PRZs as new information comes to light (e.g. on plume behaviour), linking into adaptive management.

How many PRZs (several options):

- The number of IRZs and PRZs may depend on the homogeneity of the ecosystem in the claim area, i.e. the more homogenous, the fewer PRZs are required;

- Number and location of PRZs may depend on surrounding claim areas and APEIs;

- Multiple PRZs may be required for multiple purposes (monitoring, protection of ecosystem structure and representative habitats, plume monitoring): this would take account of natural variability, allow for adaptation when one PRZ does not fulfil its function, and allow for ground-truthing of models;

- Is a plume-IRZ required? Probably not effective – a radial monitoring from the mining area would be much more efficient as the dispersion of the plume is dependent on current direction and may be quite erratic;

- A range of views were expressed on the numbers of PRZs required, ranging from "as many as possible" to "1-3" to "number as low as possible for practicality reasons", to "to be decided on a case-by-case basis depending on bio-region", and down to "zero, and more APEIs instead".

Size: Should IRZs and PRZs have the same size, and what size is optimal?

- May be determined by function (monitoring [small] vs. protection [large]);

- Large number of small PRZs vs. smaller number of larger ones: what is more effective?

- Distance between IRZ and PRZ still poorly constrained but should be constrained by expected plume dispersion;

- The size of the PRZ should be should depend on that of the impacted area.

Other issues/questions:

- Can PRZs be shared between neighbours (transboundary PRZs)?

- How do PRZs relate to EBSAs / management plans?

- Transparency in data management is important.

- Network of smaller ISA protected areas between license areas would be useful.

- Can APEIs take over the role of PRZs in particular circumstances?

- Will new selection criteria have consequences for contractors that have already defined PRZs?

- A view was expressed indicating that both IRZs and PRZs need to continue to exist throughout the mining phase and beyond closure for monitoring purposes, and PRZs perhaps in perpetuity.

Roles, responsibilities, institutional requirements and insertion of spatial management into the Draft Environmental Regulations

Spatial management was seen to be relevant for making strategic decisions for regional planning as well as for project-level planning, and was considered to be part of best environmental practice. It aids the implementation of the mandate of the ISA to protect and preserve the marine environment as expressed in Article 145 of UNCLOS. In particular, it provides spatial units for the interpretation of scientific research, monitoring and resulting measures, including cumulative and transboundary assessment of mining impacts.

More specifically, the argument was raised that APEIs on a regional level should not only have the function of being no-mining areas, but should also be used to preserve ecologically representative areas. In addition, PRZs monitored within the project site are needed for assessment of the negative effects of plumes in comparison to the IRZs. Finally, the designation of PRZs should be permanent to allow for a long-term monitoring of areas even after the closure of the mining operations.

Concerning responsibilities and institutional arrangements, it was suggested that it should be the task of the ISA to coordinate and ensure coherent planning, including a regional perspective covering APEIs, PRZs and IRZs. However, the collection of data from APEIs was seen as a challenge. A need was expressed for appropriate legal requirements and definitions of PRZs and IRZs in the future Mining Code. It was suggested that the provision(s) should also ensure review and adaptation of spatial management plans to new developments and increased knowledge, and that they should enable an integrated approach to protect vulnerable marine ecosystems.

In this regard, it was noted that sponsoring States should also play a role in assessing and reviewing data and information arising from spatial management practices. There should be transparency in deliberations on reporting and access to data (e.g. for monitoring purposes). It was suggested that monitoring and the review of environmental management plans could be outsourced to expert groups, however this raised questions concerning funding.

It was recommended that cooperation between contractors could be beneficial for an effective spatial management.

Spatial management instruments require the same procedural standards with regard to access to information, transparency, accountability, and public participation as all other ISA procedures (e.g. the EIA).

7.7 Overarching, Long-Term Environmental Strategy

7.7.1 Presentation: An Environmental Strategy?
Aline Jaeckel

This presentation focused on what an environmental management strategy for the Area could entail. Such a strategy would likely be a high-level document, setting out policies for the regulation and management of the whole Area in order to meet the requirements of Article 145 of the UN Convention on the Law of the Sea.

This policy-level document would inform regional-level and project-level management (see Figure 8). In this respect, the presentation built on the discussions of the previous day about best-practice environmental management approaches, which entail a global strategic environmental assessment (SEA) leading to a strategic environmental management plan (SEMP), as well as regional environmental assessments (REAs) leading to regional environmental management plans (REMPs). Rather that adding a layer to this best-practice model, an environmental management strategy may in effect function as a form of SEMP. Indeed, the development of a policy-level environmental management strategy could incorporate some of the elements of an SEA, such as the documentation of all potential elements of a strategy that were considered and an explanation as to why particular ones were selected.

Regarding the need for an environmental management strategy, the presentation noted four arguments. First, there are current gaps in the substantive requirements applicable to the ISA and contractors, such as the establishment of regional environmental management plans and the consideration of climate change and the full range of ecosystem services when assessing, regulating, and managing environmental impacts. Second, some existing substantive requirements are not adequately integrated into the ISA's decision-making procedure. Examples include the obligations to: apply a precautionary approach[42], prevent 'serious harmful effects on vulnerable marine ecosystems' from exploration activities[43], establish environmental baselines[44], and take into account cumulative effects of deep seabed mining and other activities when regulating and managing activities in the Area[45]. Third, moving beyond ad-hoc environmental protection measures would create regulatory certainty, enable a long-term focus of the management of the Area, and safeguard against environmental protection measures being overlooked if commercial pressure to promptly commence the exploitation phase increases. Fourth, an environmental management strategy could clarify the role of each actor (various ISA organs, sponsoring States, and contractors) in ensuring effective protection for the marine environment.

Rather than presenting conclusive findings, the presentation summarised various options for the potential content and format of an environmental management strategy, as outlined below. Importantly, some potential elements of such a strategy already exist and could be integrated into a strategic document together with

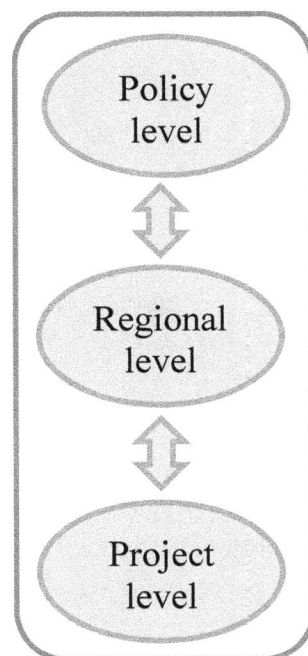

Figure 8: *Hierarchy of management levels.*

---

42 *Nodules Exploration Regulations*, regulations 2(2), 5(1), 31(2), annex IV section 5.1; *Sulphides* and *Crusts Exploration Regulations*, regulations 2(2), 5(1), 33(2), annex IV section 5.1.

43 *Nodules Exploration Regulations*, regulation 31(4); *Sulphides* and *Crusts Exploration Regulations*, regulation 33(4).

44 1994 *Agreement Relating to the Implementation of Part XI of the United Nations Convention on the Law of the Sea*, annex section 1(7); *Nodules Exploration Regulations*, regulations 18(b), 32, annex IV section 5.2; *Sulphides* and *Crusts Exploration Regulations*, regulations 20(1)(b), 34, annex IV section 5.2; ISA, *Recommendations for the guidance of contractors for the assessment of the possible environmental impacts arising from exploration for marine minerals in the Area*, ISBA/19/LTC/8 (1 March 2013).

45 ISA, *Recommendations for the guidance of contractors for the assessment of the possible environmental impacts arising from exploration for marine minerals in the Area*, ISBA/19/LTC/8 (1 March 2013), paragraph 16; ISA, *Environmental Management Plan for the Clarion-Clipperton Zone*, ISBA/17/LTC/7 (13 July 2011) paragraphs 37, 40, 51.

elements that are currently missing. The legal basis for the ISA to develop an environmental management strategy has been detailed elsewhere[46].

The following potential aims of an environmental management strategy were presented:

- To ensure all relevant environmental standards and measures are identified in a systematic manner and allocated to the appropriate actors;

- To ensure strategic environmental management is fully integrated into the Mining Code as well as the ISA's decision-making processes and supported by institutional capacity;

- To ensure environmental management measures are given effect in a timely manner.

To this end, the presentation summarised a number of potential elements of an environmental management strategy. For example, the strategy could set out overarching environmental objectives, which could guide the choice of management measures at the global, regional, and project level and be supplemented by regional and project-specific objectives.

Furthermore, the strategy could include criteria for REAs to achieve a degree of coherence between the various region-specific environmental assessments. This may also include information on how cumulative effects of mining and other activities may be taken into account.

Additionally, the strategy could outline the environmental measures required before, during, and potentially after mining. While a degree of flexibility is desirable in the management of new frontier activities, such as deep seabed mining, a certain chronology of measures can ensure regulatory certainty and methodical management. For example, if regional environmental management plans are only developed after mining has commenced within a particular region, their value is greatly diminished. To that effect, a strategic 'timeline' of steps can ensure environmental measures are procedurally integrated. Such a 'timeline' could also specify minimum data requirements needed in order to proceed to the next step.

Further elements of an environmental management strategy could be: a stakeholder engagement strategy, which is already being developed at present; an outline of how key principles, such as the precautionary approach, ecosystem approach, and best environmental practices, will be operationalised in the Area context; an outline of how the Area regime may contribute to the UN Sustainable Development Goal 14[47]; information about cooperation between the ISA and other relevant organisations in the context of Goal 14; and an allocation of specific elements of environmental measures to particular entities, such as the Secretariat of the ISA, sponsoring States, etc. The latter would ensure that institutional capacity gaps can be identified and addressed in a timely manner.

The format of an environmental management strategy could be a stand-alone policy document, which can be updated regularly and applies equally to the ISA, all member States, and contractors. Alternatively, such a strategy could be incorporated into ISA regulations, either as a separate set of regulations or as part of the future Environmental Regulations. The benefit of regulations is their unequivocally binding character, while

---

[46]   Aline Jaeckel, 'An Environmental Management Strategy for the International Seabed Authority? The Legal Basis' (2015) 30 International Journal of Marine and Coastal Law 93–119. doi: 10.1163/15718085-12341340.
[47]   Goal 14 is to 'conserve and sustainably use the oceans, seas and marine resources for sustainable development.

the disadvantage is that they are difficult to update and that any updates would not automatically apply to existing contractors.

Similarly, a crucial question relates to who should develop an environmental management strategy. Options include: the ISA Secretariat, which could provide impartial suggestions but lacks the power to adopt such a strategy; the LTC, which is a technical body instead of a policy body; or a working group of member States in the Council or Assembly, either alone or supported by consultants or participants of a workshop during which particular elements of a strategy may be discussed.

In summary, an environmental management strategy could be established to ensure the systematic and methodical development and implementation of measures to protect the marine environment from harmful effects of deep seabed mining, in order to meet the obligations outlined in Article 145 of the UN Convention on the Law of the Sea.

7.7.2   Plenary Discussion on an Overarching, Long-Term Environmental Strategy

It was asked how similar or different the proposed top-level, overarching environmental strategy would be to the strategic assessment and environmental management plans proposed as part of the tiered approach to ISA governance. Whereas similar at a global SEA level, a policy document such as the one described here would not require a fully-fletched SEA but could apply some of the methodologies and conceptual instruments of an SEA.

How would the strategy link to the lower tiers? As outlined during the sessions on regional governance (CHAPTER 7.6), strategic approaches create initiatives and determine broad goals with a means to achieving these, which can then be implemented on the regional and project level. The environmental strategy could seek to ensure coherence between regional environmental management plans for different regions.

There are different concepts for strategic documents under other international regimes. Whereas, for example, the London Protocol is quite short and concise and focuses on the long-term objectives of the regime, the Arctic Environmental Protection Strategy follows a more holistic approach, which includes objectives, instruments, and procedural requirements.

The debate then focused on the pros and cons of having provisions on strategic planning in the ISA Environmental Regulations. As the strategy level is more general by nature than regulations at the project level, it could be reasonable to have such a policy document as a living document which can be amended easily. There are elements and certain regulatory aspects which will be useful to integrate into the regulations - such as the tiers between the strategy, regional, and project level and procedural requirements.

A comment was made in the sense that one of the main challenges is to find the right balance between the different levels. Trying to clarify too many details, at the highest level and at a too early stage, for example, has the danger of blocking the whole process. The strategic document should not be only be aspirational, but at the same time, not too prescriptive.

Finally, it was said that one important benefit of a strategic document would be that it allows for consultation with other international organisations in charge of other uses of the marine environment.

7.7.3   World Café Discussions

Environmental strategy – Who? How? By When?

Is an environmental strategy necessary?

There were different views on the fundamental question of whether an environmental strategy is necessary, and if so, why. On this issue, there was also the view that there is a need for a higher level, self-standing strategic or policy document. Such a document could contain goals, environmental objectives, and mechanisms to operationalise the precautionary approach, ecosystem-based approach and the common heritage of mankind principle in a systematic and overarching way. This document could feed into, be connected to support, and set the context for, regional and contractor-specific Environmental Regulations. However, the level of prescriptiveness (normative vs. aspirational) must be decided upon in advance: ideally, the environmental strategy should be specific enough to provide real guidance, but not so detailed that it runs the risk of not being adopted:

- A clear distinction needs to be made between an SEA and the environmental strategy to avoid duplication. Is there a way to merge the two so that the environmental strategy builds on the SEA and in effect becomes the global Strategic Environmental Management Plan (SEMP)?

- An ISA global-level environmental strategy could be beneficial irrespective of the fact that lower level regulations and management tools already exist;

- UNCLOS already provides the elements for an overarching environmental strategy. The further development of the strategy should get urgent attention as industry is already planning and monitoring is ongoing;

- An environmental strategy would support coherence in the planning process (data gathering vs. conservation objectives vs. use [MPAs, APEIs, and mining sites]);

- An environmental strategy should not, however, be used as a substitute for binding regulations where these are needed.

What should the environmental strategy contain?

An environmental strategy which sets out global environmental objectives and standards could:

- Help to inform the definition of harmful effects which would, in turn, assist contractors in the development of EIAs and risk assessments;

- Be an overarching document that incorporates the findings of the SEA and informs, or is in fact, the SEMP (which ideally goes all the way to EIA level);

- Build on the lessons and principles from the CCZ-EMP and/or the Arctic Strategic Plan;

- Inform other regional strategies for areas of ocean with no strategy yet (criteria for REA; separate REMP for each mineral type?);

- Integrate the precautionary approach and the common heritage of mankind as basis;

- Help operationalise the precautionary approach;

- Incorporate global climate modelling and predictions;

- Encourage marine scientific research;

- Include a list of obligations (e.g. to carry out an REA, transparency), roles (differentiating between ISA, sponsoring States and contractors) and responsibilities;

- Set the frame for specific regulations.

Who should develop a global environmental strategy?

- The ISA should clearly lead, but will need to strengthen its capacity to do so. Nevertheless, an effective environmental strategy is multi-sectoral: minerals cannot be seen in isolation;

- The ISA Secretariat needs to be tasked with, and mandated by the Assembly/Council to develop the strategy/policy (as the Assembly is in charge of overall policies for the ISA), or alternatively States Parties could call on the Assembly to develop an environmental strategy => which could then be developed by the Secretariat and its organs with broad engagement by all stakeholders (including scientists, international organisations, etc. => "independent expert advice");

- Policy may identify the need for a new body, which can then spawn regulations. The view was expressed that this would be a job for a new Environmental Committee that could, inter alia, (1) implement the strategy, (2) underline the tools required for SEA, REA, EIA, and (3) stand in open dialogue with stakeholders to make the development of the environmental strategy a systematic and formal process;

- There was some discussion about "interested persons" and who should be consulted, including communities from States that could be affected by operations. It was recognised that this was probably not so much of an issue for the CCZ but could be for other regions such as the Azores where there are SMS sites relatively close to shore. Comments were made that this consultation could be covered through, for example, the Assembly where Portugal is a Member State (Portugal can also be an observer to Council and provide input there). It was also mentioned that civil society is represented through observers, noting that NGOs/civil society groups have begun forming coalitions to ensure the wider civil society voice is heard;

- It was noted that it is not normal practice to put out policies for public consultation – so it would be welcomed but not expected;

- The question was raised; could the UN General Assembly call upon the ISA to develop such a strategy?

- Evidence of the past shows that strategies are most effective if developed by those that eventually implement them. Also, it was noted that good policy is generally not written by lawyers or scientists but rather by policy experts;

- The strategy could form a "chapter" of an overall strategic plan for the development of the ISA, or could even be included as an attachment to the Environmental Regulations.

How should the environmental strategy be developed?

- Will require an increase in capacities within the ISA, e.g. environmental, technical, engineering;

- Comments were made in the sense that it should be done in close cooperation with and/or by developing MoUs with other global organisations (e.g. FAO, IMO) and should build on existing (regional) frameworks of other regimes (e.g. CBD, RFMOs, RSCs, UNESCO-IOC, LC/LP). Research efforts should also be coordinated ("collective arrangement"), but the strategy needs to be tailor-made for and formalised within the ISA;

- Many basic environmental strategic elements, including principles and legally binding requirements to inform an environmental strategy already exist in legally binding and highly influential legal instruments (e.g. UNCLOS/IA, ISA Regulations/Recommendations/Guidelines, Discussion Paper on Environmental Regulations, LTC documents, ITLOS Advisory Opinion). A proposal was made to compile those in a single document as a basic underpinning source;

- A range of opinions were made, including workshops and, consultants, but also input from science and the wider stakeholder base is required (e.g. stakeholder submissions before drafting policy documents);

- Sufficient data must be made available;

- Document the process in academic literature (e.g. Wedding et al., 2015 for the CCZ);

- A view considered that non-sponsoring States should be involved to ensure independence;

- The strategy should not be developed in isolation from the UN Preparatory Committee talks on a new treaty for Marine Biodiversity Beyond National Jurisdiction (BBNJ);

- One WC group proposed the need for an initial "non-paper" - i.e. non-attributable approach to a written document as occurred in original UNCLOS negotiations to deal with intractable issues. In this context, there could be several "non-papers" setting out the authors' view of an ideal environmental strategy, including a flow chart if possible (e.g. from ENGOs, contractors, any other grouping of individual stakeholders, etc.). The authors, or even the groups, can be anonymous. According to this idea, these would be reviewed by the Secretariat and its organs to see where they overlap, where they conflict, or where there are gaps, and the output from the non-paper approach would be a compendium of a draft environmental strategy as the basis for further focused discussion.

By when should an environmental strategy be in place?

- The opinion was expressed that in an ideal world, the development of an environmental strategy would precede other documents, but that this has not been the case. Nevertheless, a view was held that it is never too late. In the light of the UNCLOS Article 154 review, this strategy was considered to be a response to some of the concerns raised;

- Taking into account the other priorities of the ISA, the strategy could be developed in parallel with and simultaneously inform the development of the Environmental Regulations, 2 to 3 years from now;

- A view was raised in the sense that the environmental strategy should be in place before any exploitation contracts are approved, and that granting of mining licenses (contracts) should be dependent on a prior adoption of an environmental strategy. This could also include a certain responsibility of contractors to collect regional baseline data beyond their contract area(s). This view was one amongst others, as well as one that generated a diversity of opinions;

- Developing an environmental strategy should not slow down the entire process and should be developed in parallel with regional/contractor-level regulations;

- The point was raised that environmental objectives are needed soon to inform lower levels and to allow contractors to calculate their costs;

- Mechanisms to update a SEMP once the REAs/EIAs have been carried out should be in place to identify flaws (adaptive management).

8. Plenary Discussion on How to Involve Science Effectively

The moderator began by analysing different sections of the Draft EnvRegs, where scientific input was considered necessary. Although it was considered that scientific input is essential, it remains to be determined whether this should come primarily from the academic community or from the various contractors. The question was then rephrased to clarify what science should provide and what function it should perform. The question raised, in particular, was "what forms of scientific input are needed in the development of Regional Environmental Plans/Assessments?"

The core competencies of science were summarised as being (1) data generation, (2) data interpretation (setting the context; integration of theory, and available data), (3) metadata-analysis, such as used for the development of the CCZ-EMP, and (4) providing expert advice based on best available knowledge, and partnering in the development of technologies, protocols, and methodologies. More specifically, science contributes to regional governance by (1) providing baseline data, (2) allowing for spatial mapping and planning processes, and (3) producing spatial databases and models of the impacts of multiple operations at the same site, as well as other cumulative and synergistic impacts. Applied science provides input into regulatory processes, conducts contract-specific research activities including monitoring, evaluates technologies, assesses best available techniques, contributes to the designation of protected areas, and contributes to determining performance standards for contractors. In addition, it identifies potential mitigation measures, sets guiding principles for the frequency and scale of monitoring activities, and provides recommendations on best practice.

A prerequisite for ensuring the engagement of science in the DSM industry and the development of mining regulations is national and international funding. National funding agencies are currently very reluctant to support scientific projects in the Area. Action is required in four particular areas to encourage future scientific activities:

- Political and social agendas need to promote deep-sea research and align national priorities with those needed for research in the Area: Science can be involved effectively through policy (e.g. G7 meeting in Germany, where an advisory note on DSM was assembled for the meeting of the science ministers);

- Research priorities need to be established, which are attractive to national research funding agencies and align with their priorities. The ISA can assist marine scientific research (MSR) by developing pertinent arguments for funding, clearly defining its scientific needs and priorities, and by involving other organisations such as IOC, Global Ocean Biodiversity Initiative, etc. and the MSR community. Developing special interest groups is also an option, e.g. the newly established GOOS section DOOS (Deep ocean observing system) could set up a sub-group of/for deep seabed mining;

- Workshops are an effective way to develop appropriate REAs and REMPs based on best available science (e.g. CCZ-EMP) and need not be costly. On the policy side, member States could invite scientists to join their delegations, so that relevant scientific experts become more involved in decision-making forums;

- A gap analysis and metadata analysis is needed to provide the "big picture" and target scientific effort. Collaboration between MSR and contractors and between contractors does already occur but should be further encouraged. National research in the Area should be integrated into the knowledge base. Communication opportunities bridging the gaps between ISA and MSR, contractors and civil society should be encouraged.

An interactive working mode between the ISA and science is required to stay informed and be aware of state-of-the-art monitoring and baseline studies. A formal structure for this is currently lacking, e.g. a Scientific Advisory Board. In order to keep the MSR community involved, it will be necessary for the ISA to ensure the freedom of marine scientific research in the Area and to establish agreed codes of conduct for working within contract areas.

Transparency of scientific results/data/interpretation is sorely needed. The MSR community is required to publish all data after the end of a project, so there is an imbalance in the requirements posed to MSR actors and consultancies when it comes to cooperation with contractors. More transparency could lead to fairer and better cooperation with science. The MSR community, the ISA and contractors should all be required to make their findings mutually accessible. Interconnected databases would provide a means for making research data accessible, including the ISA's data management practices that require expansion.

A time-line could be developed that highlights specific phases during which scientific information is needed during the development of the Exploitation Regulations and during the mining process. This should indicate whether scientific advice or information is needed regularly or only on the request of the ISA (e.g. update of best available practice).

9. Side Event: Role and Function of the Sponsoring State

9.1 Singapore Legislation on the Role and Function of the Sponsoring State
Rena Lee

The Deep Seabed Mining Act of Singapore came into effect on 1 April 2015. In drafting the Act, Singapore had reference to UNCLOS and the Part XI Agreement, as well as the Advisory Opinion of the Seabed Disputes Chamber of 2011 on the responsibilities and obligations of sponsoring States.

The enactment of the Deep Seabed Mining Act fulfils Singapore's "due diligence" obligation as a sponsoring State. Pursuant to Article 139(2) and Article 4(4) of Annex III of UNCLOS, the enactment of legislation is one of the "reasonably appropriate" measures that sponsoring States should take.

The purposes of the Act, as set out in Section 3, are to regulate activities by the sponsored entities, ensure effective protection of the marine environment from harmful effects of exploration and exploitation activities, and fulfil Singapore's obligations in relation to activities in the Area.

The Act establishes a licensing regime to regulate the exploration and exploitation activities of sponsored entities. Any Singapore company wishing to undertake such activities must apply to the Singapore Government for a licence. This licence is only granted if the Government is satisfied that the applicant meets or will meet the qualification standards in Article 4 of Annex III of UNCLOS and that it intends to apply to the ISA for a contract. Only then can an applicant obtain a certificate of sponsorship.

In granting the licence, the Government may subject the licence-holder to conditions, which include, among others, the requirement to comply with relevant provisions of UNCLOS, the Part XI Agreement, the rules, regulations and procedures of the ISA, the ISA's decisions as well as the terms of its contract with the ISA. Such conditions reflect the key rationale for the system of sponsorship by States Parties.

In the event that the licence holder breaches any condition of its licence, he is liable to a financial penalty not exceeding S$40,000. The Government may also suspend or revoke the license. Apart from the licensing regime, the Act also provides for the enforcement of decisions of the Seabed Disputes Chamber and arbitral awards pursuant to Article 188(2)(a) of UNCLOS.

While sponsoring States have a role to play in regulating the activities of their sponsored activities, the primary responsibility for regulating conduct in the Area belongs to the ISA. Sponsoring States are obliged to assist the ISA, pursuant to Article 153(4) of UNCLOS. Looking ahead, greater collaboration should be fostered, especially as States, such as flag States and coastal States, also become involved.

9.2 Chinese Legislation on the Role and Function of the Sponsoring State
Xuewei Xu

Activities in the Area are entering a critical stage, with the focus shifting from exploration to exploitation. Regulations for the Area are a significant component in determining the obligations and exemption from liability for sponsoring States. The Standing Commission of the National People's Congress of the People's

Republic of China (PRC) adopted the Law on Resource Exploration and Exploitation in the Deep Seabed Area on February 26, 2016 after a long period of preparation and a significant increase in activities in the Area. This law entered into force on May 1, 2016. In general, the law was enacted for the purpose of regulating the activities of resource exploration and exploitation in the deep seabed area, advancing science and technology research, investigating resources, protecting the marine environment, promoting the sustainable utilisation of deep seabed resources and maintaining the common interests of mankind. It consists of 7 chapters and 29 articles addressing general provisions, exploration and exploitation, environmental protection, science and technology research and resource investigation, supervision and inspection, legal liability, and supplementary provisions. It should be noted that drafting a law for the deep seabed is never an easy task and the relevant rules and regulations will continue to be built on and improved.

9.3 Activities in the Area and the Role of Sponsoring States: An Institutional Perspective
Pradeep Singh

The presentation began with a concise description of the concept of State sponsorship, explaining what it is and why it is important in the context of mining activities in the Area. Thereafter, the role of sponsoring States with respect to the protection of the marine environment was discussed. Particular attention was given to the 2011 Advisory Opinion on the responsibilities and obligations of States sponsoring persons and entities with respect to activities in the Area (ITLOS Case No. 17) in this regard. By referring to the obligations of states to protect the marine environment from harmful activities conducted under their control or permission, a submission was made that these obligations under Part XII of UNCLOS and customary international law may be extended to Part XI and the international seabed. Accordingly, the presenter contended that sponsoring States are not only obliged under international law to ensure that national laws are in place in their domestic legal system to ensure the conformity of their sponsored entities with regulations, guidelines, and contractual obligations as well as to assist the ISA in ensuring compliance (and enforcement where necessary), but are also further obligated to take proactive measures such as to conduct environmental assessments, participate in the preparation of project-specific environmental management plans, and most crucially to continuously monitor and report on the environmental impacts of activities carried out by their sponsored entities.

Despite this, it was highlighted that there currently is no clear division of responsibilities and tasks between the ISA, sponsoring States, and contractors. The deficiencies in the current institutional structure were pointed out (such as the lack of rules or guidelines to regulate their conduct and the absence of a common platform for interaction), but it was also acknowledged that steps are being taken to address some of these concerns. The presentation ended with several points for discussion, including: (1) how should the relationship between sponsoring States and the ISA be finalised (i.e. should it be binding or non-binding); (2) what should the scope of the content be (i.e. how detailed or prescriptive should it be, as well as frequency of action; (3) what standards and outcomes should be expected for monitoring activities, given that sponsoring States have different capacities and abilities; (4) the need to include independent scientific research agencies in the monitoring regime; (5) the division of responsibilities in instances where an entity has more than one sponsoring State; (6) the possibility of a conflict of interest situation arising if a sponsoring State is part of an inspectorate to inspect the performance of an entity which it sponsors; and (7) whether there is a need for a follow-up advisory opinion to clarify some of these doubts. In conclusion, it was stressed that sponsoring

States can and must play a crucial role in the carrying out of activities in the Area and that there is an urgent need for an expert workshop to be held in the near future to address these topics.

9.4   Summary of Discussions During the Side Event; Role and Function of the Sponsoring State

It was firstly clarified that Part XII should be considered when defining the obligations of the ISA with regard to the protection of the marine environment under Part XI. The ISA is formed by the Contracting States Parties to UNCLOS on the one hand, but also consists of its various organs such as the Assembly, the Council and the Legal and Technical Commission, as well as the Secretariat. These organs have different roles with regard to the protection of the marine environment according to UNCLOS.

A second point of the discussion referred to Article 21 of Annex III of UNCLOS, which allows for more stringent requirements to be put in place by States Parties and for the application of national environmental rules and provisions to vessels flying their flags. It was asked whether this is already the case according to the national provisions presented during the side event. For Singapore, the national rules are flexible enough to put in place more stringent requirements if this is adequate. It was mentioned that this fits with the due diligence obligation outlined by the ITLOS advisory opinion. So far, no need has occurred – but that might change if exploitation activities start. The same is true with regard to vessels flying the flag of Singapore.

Similarly, no more stringent requirements on EIA have been put in place in Singapore or China. The point was raised that imposing more stringent requirements may trigger juridical action. In any case, an appropriate justification for more stringent requirements seems to be required.

Thirdly, it was questioned whether the national laws presented foresee clear consequences for a breach of the direct obligations of the sponsoring State, such as the implementation of the precautionary approach or the application of Best Environmental Practice. It was clarified that no breach could be documented as long as there are not clear environmental thresholds or obligations in place. Moreover, the consequence of a breach of a direct obligation must also take into account the actual harm which has been or which might have been caused. The relevance might again increase in the case of future exploitation activities.

10. NGOs Statements

10.1 Seas At Risk
Ann Dom

Seas At Risk is an umbrella organisation of 34 environmental NGOs from across Europe that promotes ambitious policies for marine protection at European and international level. Seas At Risk's 'Deep sea mining? Stop and think!' position is that deep seabed mining has no place in the world's Agenda 2030 for sustainable development, and that the UN and EU should focus their policies on sustainable resource use and production instead. With socio-economic benefits that are bound to be short-term (and are still highly uncertain), and the risk of irreversible and significant environmental impacts, deep seabed mining poses a serious threat to sustainability. The precautionary principle advises to prioritise sustainable alternatives to avoid our economy becoming locked into this high-risk technology. Furthermore, with the world committed to Agenda 2030, economies will have to transition to much more resource-efficient systems and life-style changes in order to meet the SDGs, in particular the goal on responsible consumption and production. This questions the future need for deep seabed mining.

Alternatives to deep seabed mining are available, and can be found in the transition of economies towards more sustainable models. This includes not only a transition to a circular economy (eco-design, repair, re-use, recycling, substitute materials etc.) but also to smart mobility and energy systems, smart cities, and new lifestyles. To date, however, such alternative options have hardly been explored. Seas At Risks calls on the UN and the EU to carry out a future outlooks study on mineral resource demand and supply under various 2050 scenarios, i.e. a business-as-usual scenario, a circular economy scenario, and a scenario with a full transition to sustainable economies. Such a study could inform the debate among stakeholders and hopefully give an objective answer to the question whether deep seabed mining is needed in order to supply our future economies with the necessary resources. It would help to underpin a wider public debate as well as inform future policy initiatives at international and EU level.

10.2 Deep Sea Conservation Coalition
Matthew Gianni

The Deep Sea Conservation Coalition (DSCC) consists of over 70 member organisations, many of whom are grappling with the issue of deep seabed mining. The position of the DSCC currently is that the consumption of mineral resources should be based on sustainability, re-use, improved product design, and recycling of materials rather than exploring for new sources of minerals, including in the deep sea. And while some proponents argue that deep seabed mining is needed to provide sufficient metals to transition to a renewable energy economy, a report released last year concluded that even the most ambitious scenario - a transition to a 100% renewable energy economy by 2050 - can be done without sourcing supplies of metals such as copper, cobalt, nickel, lithium, silver, and specialty metals such as tellurium and rare earth metals from the deep sea[48].

If deep seabed mining is permitted to occur, the DSCC position is that the full range of marine habitats, biodiversity, and ecosystem functions need to be adequately and effectively protected, through instalment of MPAs/APEIs, the development of regional environment management plans, collection of robust and adequate baseline information, the precautionary approach, applying the polluter pays principle, and establishing a liability fund and a sustainability fund, amongst other measures. The ISA must become much more transparent, particularly the LTC, and an environment committee is needed. However, even with the best regulations in place, many NGOs have major concerns as to whether deep seabed mining can ever be managed with minimal environmental impact given the uncertainties about biodiversity and ecosystem function in the deep sea, the difficulties in monitoring impacts of mining activities, and the potential for long-term and irreversible harm. Additional concerns centre on whether the political, institutional, and financial structures can be put in place to ensure effective monitoring of mining activities and compliance with environmental regulations.

A lot is at stake. The UN's 1st World Ocean Assessment in 2015 states that the deep-sea "constitutes the largest source of species and ecosystem diversity on Earth", which supports diverse ecosystem processes and functions necessary for the Earth's natural systems to function and yet is already under threat from multiple stressors[49]. And States have committed, in Sustainable Development Goal 14.2, to, by 2020, sustainably manage and protect marine and coastal ecosystems to avoid significant adverse impacts, including by strengthening their resilience, and taking action for their restoration in order to achieve healthy and productive oceans[50]. Can deep seabed mining be compatible with these concerns and objectives?

10.3   Greenpeace
Kathryn Miller

Greenpeace takes the view that no seabed mining applications should be granted, and that no exploration or exploitation should take place, unless and until the full range of marine habitats, biodiversity, and ecosystem functions are adequately protected[51]. In order to contribute towards a precautionary system of regulation, which can ensure the protection of species and habitats and the future provision of ecosystem services and resources, Greenpeace stresses the urgency of:

- Greater transparency in the ISA, including that the Legal and Technical Commission follow international practice and is open to observers. We welcome the Article 154 committee report in that regard;

- Public release of environmental data, including those from baseline surveys and monitoring;

---

[48]   Teske, S., Florin, N., Dominish, E. & Giurco, D. 2016, Renewable Energy and Deep Sea Mining: Supply, Demand and Scenarios. University of Technology Sydney https://opus.lib.uts.edu.au/handle/10453/67336

[49]   UN World Oceans Assessment. 2015. Chapter 36F. Open Ocean and Deep-Sea. http://www.worldoceanassessment.org/?page_id=14

[50]   http://www.un.org/sustainabledevelopment/oceans/

[51]   Greenpeace is campaigning for the implementation of a global network of Marine Reserves that would protect at least 40% of the world's oceans, including particularly vulnerable areas such as seamounts, from threats such as seabed mining.

- More effective engagement between the ISA and other international bodies and organisations with relevant expertise and experience, including IMO and the London Convention and Protocol, as well as with NGOs and others, in order to work towards standardisation of guidance and regulations;

- Action to address the current mismatch and lack of harmonisation between regulations relating to the ABNJ and EEZs.

10.4   World Wildlife Fund
Simon Walmsley

WWF believes that deep seabed mining activities should not commence before measures are in place to protect deep-sea ecosystems from adverse impacts, and then only overseen by an equitable governance system that has completed a series of steps outlined below.

Responsible States and the ISA, in bio-regions being considered for seabed exploration or mining, have established an equitable governance system that has accomplished the following steps:

- Openly and transparently considered alternatives to mining deep-sea minerals, taking into account ecological, social and economic perspectives, including: conserving natural and mineral resources; increasing the recycling of minerals; and exploiting land-based mineral resources with much greater efficiency and more stringent environmental regulation;

- Carried out strategic environmental assessments of the likely impacts of deep seabed mining on the marine environment, including the potential cumulative effects in conjunction with other human activities;

- Prepared and implemented ecosystem-based oceans management strategies, laws, and regulations that: collect adequate baseline information on the marine environment where mining could potentially occur, including the location of sensitive deep-sea habitats/ecosystems; establish a comprehensive network of well-managed protected areas to protect vulnerable marine ecosystems, ecologically or biologically significant areas, depleted, threatened or endangered species, and representative examples of deep-sea ecosystems;

- Adopted a precautionary approach that assumes that deep seabed mining will have adverse ecological impacts in the absence of compelling evidence to the contrary;

- Defined standards for the environment around any deep-sea operation, building on local, national and regional knowledge of the sensitivities of deep-sea ecosystems, to minimise environmental impacts and avoid significant and irreversible adverse environmental impacts;

- Permitted exploration or exploitation of minerals on or below the seabed only following Environmental Impact Assessments for each potential project that include full identification, assessment, and treatment of risks (including those with low probability, but high consequence);

- Assigned liability to the owners or operators of exploration or exploitation facilities for the costs associated with the containment or clean-up of any unauthorised discharges of materials and/or waste, and any damages resulting from such discharges ("polluter pays");

- Established contractor-independent public assessment and monitoring of the permit conditions and potentially impacted ecosystems; and

- Established a comprehensive and adequately-funded mechanism to cover clean-up costs, damages to affected parties, and the restoration of the environment associated with unauthorised discharges of materials and/or waste where the responsible party is unknown, unable to pay, or refuses to pay.

11. Appendices

11.1 Workshop Programme

Monday 20 March 2017

Time	Room[52]	Presenter / *Moderator*	Activity
9.00 - 9.10	PL	Lilian Busse	Introduction: Objective of workshop + Workshop modalities
9.10 – 9.20	PL	Ralph Watzel, BGR	Welcome
9.20 – 9.30	PL	Lilian Busse, UBA	Welcome
9.30 – 9.45	PL	Michael Lodge, ISA	Welcome and Background to ISA mandate and regulations
9.45 – 9.50	PL		Introduction of speakers
9.50 – 10.15	PL	Antje Boetius	Talk on potential impacts of exploitation activities on the marine environment
10.15 – 10.40	PL	Lisa Levin	Talk on which levels of „harm" caused by deep seabed mining are acceptable from a scientific perspective
10.40 – 11.00	PL		Discussion
11.00 – 11.30		Coffee	

---

[52] PL = plenary

Time	Room[52]	Presenter / Moderator	Activity
11.30 – 12.10	PL	Christopher Brown	Talk: Overview of the ISA Draft Env. Regulations
12.10 – 12.20	PL	Christian Reichert	Critical Statement on Draft Env. Regulations from the LTC
12.20 – 12.30	PL	Duncan Currie	Critical Statement on Draft Env. Regulations from a legal science perspective
12.30 – 12.40	PL	Eva Ramirez-Llodra	Critical Statement on Draft Env. Regulations from a natural science perspective
12.40 – 12.50	PL	Ralph Spickermann	Critical Statement on Draft Env. Regulations from a contractor perspective
12.50 – 13.00	PL	Harald Ginzky	Critical Statement on Draft Env. Regulations from the German Environmental Agency
13.00 – 14.00			Lunch
14.00 – 14.10	PL		Explanation World Café discussions. Participants will be sent to three rooms (approx. 30 per room), each with three discussion tables.
14.10 – 16.00		*Q1: Sebastian Unger, Andrew Birchenough, Ralph Spickermann* *Q2: Chris Brown, Carsten Rühlemann, Kristina Gjerde* *Q3: Pradeep Singh, Sabine Christiansen, Sven Mißling*	World Café discussions: Question 1: Is the structure and content of the working draft adequate / fit-for-purpose? Question 2: Is the working draft too prescriptive or not? Question 3: Gap analysis of the overarching objectives and the strategic approach

Time	Room[52]	Presenter / *Moderator*	Activity
16.00 – 16.30		Coffee	
16.30 – 17.00	PL	World Café moderators	Feedback of discussions on the three World Café questions
17.00 – 17.20	PL	Dave Billett	Talk: PEW analysis of the Draft Env. Regulations
17.20 – 18.00	PL		Discussion
18.00			End of day 1

Tuesday 21 March 2017

Time	Room	Presenter / *Moderator*	Activity
		TOPIC SUBSTANTIVE CRITERIA	
9.00 – 9.25	PL	Robin Warner	Talk: Common understanding of relevant principles of international environmental law (in view of the common heritage of mankind-principle) and their interplay
9.25 – 9.45	PL		Discussion
9.45 – 10.10	PL	Robin Warner	Talk: Clarification of legal threshold criteria ("harm", "serious harm" and/or "effective protection of the marine environment from harmful effects")
10.10 – 10.30	PL		Discussion

Time	Room	Presenter / *Moderator*	Activity
10.30			Coffee
		TOPIC ENVIRONMENTAL STANDARDS	
11.00 – 11.25	PL	Christopher Brown	Talk: The role of standards in the ISA Draft Env. Regulations
11.25 – 11.45	PL	Sabine Christiansen	Talk: MIDAS results on available environmental standards
11.45 - 12.05	PL	Roland Cormier	Talk: UNECE Standard for risk management in regulatory frameworks and the implementation of the ecosystem approach to management
12.05 – 12.10	PL		Introduction of parallel working groups (3 rooms with ca. 30 participants per room)
12.10 – 12.40		*Q1: Samantha Smith, Annemiek Vink*	Group discussions: Question 1: Develop working methodology to create ISA standards based on existing standards and guidance
		Q2: Torsten Thiele, Katherine Houghton	Question 2: Use of standards: Compulsory vs. voluntary? Pros and cons
		Q3: Roland Cormier, Stefan Bräger	Question 3: Using risk management for the implementation of the ecosystem approach to managing DSM
12.40 – 13.00	PL	Working group rapporteurs	Feedback from the three working groups
13.00 – 14.00			Lunch

TOPIC EIA/EIS

14.00 – 14.40	PL	Malcolm Clark	Talk: EIA state-of-the-art development, incl. finalisation of draft templates EIA/EIS and clarification of overall EIS contents
14.40 – 14.45	PL		Introduction of parallel working groups
14.45 – 15.30			Group discussions:
		Q1: Malcolm Clark	Question 1: finalisation of EIA/EIS templates
		Q2: Neil Craik, David Billet	Question 2: EIA process - roles and responsibilities
		Q3: Thomas Merck, Katherine Houghton	Question 3: Standardisation of assessments and monitoring
15.30 – 16.00	PL	Working group rapporteurs	Feedback from the three working groups
16.00 – 16.30			Coffee
16.30 – 16.35	PL		Introduction of World Café discussions
16.35 – 17.35			World Café discussions:
		Q1: Adrian Flynn, Kristina Gjerde, Lisa Levin	Question 1: When are environmental impacts significant, how can these be determined and which consequences do these have?
		Q2: Se-Jong Ju, Wini Broadbelt, Simon Walmsley	Question 2: How to deal with uncertainties
		Q3: Jeff Ardron, Uwe Jenisch, Shaojun Liu	Question 3: Consequences of the EIA for decision-making
17.35 – 18.00	PL	World Café moderators	Feedback of discussions on the three World Café questions
18.00			end of sessions day 2

SIDE EVENT: ROLE AND FUNCTION OF SPONSORING STATE			
19.00 – 19.10	PL	Philomène Verlaan	Introduction
19.10 – 19.25	PL	Rena Lee	Talk: Singapore Legislation on the role and function of sponsoring State
19.25 – 19.35	PL	Xuewei Xu	Talk: Chinese Legislation on the role and function of sponsoring State
19.35 – 19.45	PL	Pradeep Singh	Talk: The role of sponsoring State from an institutional perspective – division of responsibilities
19.45 – 20.00	PL	Philomène Verlaan	Discussion
20.15	WORKSHOP DINNER (Hotel restaurant "Alte Meierei")		

Wednesday, 22 March 2017

Time	Room	Presenter / *Moderator*	Activity
TOPIC ADAPTIVE GOVERNANCE			
9.00 – 9.30	PL	Neil Craik	Talk: Incorporating flexibility into environmental governance with regard to the Env. Code
9.30 - 9.40	PL	Guifang Xue	Discussion
9.40 – 9.45	PL		Introduction of parallel working groups

Time	Room	Presenter / *Moderator*	Activity
9.45 – 10.30			<u>Working group discussions:</u>
		Q1: Neil Craik, Arne Küper	Question 1: Is the deep-seabed mining regime suitable for the use of adaptive governance and management?
		Q2: Aline Jaeckel, Matt Gianni	Question 2: Options for an adaptive regulatory framework to be incorporated into the regulations
		Q3: Kris van Nijen, Philomène Verlaan	Question 3: Opportunities and obligations of contractors to adapt their mining operations after the contract is concluded (e.g. through innovation/ development of BAT and BEP)
10.30 – 11.00	PL	Working group rapporteurs	Feedback from the three working groups
11.00 – 11.30			Coffee

TOPIC PILOT MINING TESTS

Time	Room	Presenter / Moderator	Activity
11.30 - 11.55	PL	Katherine Houghton	Talk: Challenges around the testing of mining equipment and processes
11.55 – 12.00	PL		Introduction of parallel working groups
12.00 – 13.00			<u>Working group discussions:</u>
		Q1: Duncan Currie, Laleta Davis Mattis	Question 1: How can Pilot Mining Tests be integrated into the regulatory process/the regulations?
		Q2: Robin Warner, Felix Janssen	Question 2: Can a multi-phase EIA process be installed for test mining?
13.00 – 13.30	PL	Working group rapporteurs	Feedback from the two working groups
13.30 – 14.30			Lunch

Time	Room	Presenter / *Moderator*	Activity
14.30 – 15.30	PL		Statements of NGOs
15:30 – 16:30		*Group 1: Malcolm Clark* *Group 2: Roland Cormier* *Group 3: Katherine Houghton* *Group 4: Jennifer Warren*	<u>Working Group discussions</u>: Group 1: Finalisation of EIA/EIS templates (cont'd from Tuesday) Group 2: Risk assessment and management (cont'd from Tuesday) Group 3: Pilot Mining Tests Group 4: incentives for contractors
16.30 – 17.00			Coffee
17.00			CITY TOUR

Thursday 23 March 2017

Time	Room	Presenter / *Moderator*	Activity
		TOPIC REGIONAL GOVERNANCE	
9.00 – 9.35	PL	Philip Weaver, Daniel Jones	Talk: Overarching issues around regional governance of DSM
9.35 – 9.40	PL		Introduction of World Café discussions

Time	Room	Presenter / *Moderator*	Activity
9.40 – 10.40		*Q1: Marta Ribeiro, Jeff Ardron, Daniel Jones* *Q2: Alfonso Ascencio-Herrera, Maria-Sofia Villanueva, Phil Weaver* *Q3: Kristina Gjerde, Amber Cobley, Sabine Gollner*	World Café discussions: Question 1: Is a multiple-tiers approach to management required? First thoughts as preparation for Friday's discussion? Question 2: What are the pros and cons of regional governance? Question 3: Objective criteria for initialising regional governance? Who is in charge?
10.40 – 11.30	PL	World Café moderators	Feedback of discussions on the three World Café questions
11.30 – 12.00			Coffee
12.00 – 13.00	PL	*Gordon Patterson, Kristin Hamann*	Plenary discussion: How to involve science effectively?
13.00 – 14.00			Lunch
14.00 – 14.30	PL	Daniel Jones, Philip Weaver	Talk: Spatial management approaches
14.30 – 14.40	PL		Introduction of four World Café discussions

Time	Room	Presenter / _Moderator_	Activity
14.40 – 16.00			<u>World Café discussions:</u>
		Q1: Phil Weaver, Ingo Narberhaus	Question 1: Identification of required spatial measures to minimise environmental impacts - in addition to the designation of APEIs?
		Q2: Daniel Jones, José Angel Perez	Question 2: Complement the required elements of the CCZ REMP to derive a REMP template
		Q3: Annemiek Vink, Stefan Bräger	Question 3: Selection of areas as PRZ and IRZs - how, how many and where?
		Q4: Harald Ginzky, Michelle Walker	Question 4: Roles, responsibilities, institutional requirements and insertion of spatial management into the draft Environmental Regulations
16.00 – 16.30			Coffee
16.30 – 17.30	PL	World Café moderators	Feedback of discussions on the four World Café questions
17.30			End of day 4

Friday 24 March 2017

Time	Room	Presenter / _Moderator_	Activity
		TOPIC OVERARCHING LONG-TERM ENVIRONMENTAL STRATEGY	
9.00 – 9.30	PL	Aline Jaeckel	Talk: Necessity and elements of an Environmental Strategy

Time	Room	Presenter / *Moderator*	Activity
9.30 – 9.40	PL		Introduction of World Café discussions, with just 1 topic in two rooms (5 tables)
9.40 – 10.40		*Kristina Gjerde, Aline Jaeckel, Samantha Smith, Sebastian Unger, Philomène Verlaan*	<u>World Café discussions:</u> Question: Environmental Strategy - Who? How? By when?
10.40 – 11.30	PL	World Café moderators	Feedback of discussions on the World Café question
11.30 – 12.00			Coffee
WRAP UP and NEXT STEPS			
12.00 – 14.00	PL		Discussion: - Synopsis of results, in particular with regard to potential amendments to the Draft Env. Regulations - Roadmap for future requirements and activities
14.00 – 15.00			Lunch
15.00 – 16.00			Final Meeting of Steering Committee
16.00			End of workshop

11.2 List of Participants

Nr.	NAME	AFFILIATION	EMAIL
1.	Ardron, Jeff	The Commonwealth Secretariat / UK	j.ardron@commonwealth.int
2.	Ascencio Herrera, Alfonso	International Seabed Authority / Jamaica	alfonsoa@isa.org.jm
3.	Arjan Singh, Pradeep	University of Bremen / Germany	pradeep@uni-bremen.de
4.	Billett, David	Deep Seas Environmental Solutions Ltd. / UK	david.billett@deepseasolutions.co.uk
5.	Birchenough, Andrew	Cefas, Lowestoft Laboratory / UK	andrew.birchenough@cefas.co.uk
6.	Boetius, Antje	Alfred-Wegener-Institute / Germany	antje.boetius@awi.de
7.	Bräger, Stefan	International Seabed Authority / Jamaica	sbrager@isa.org.jm
8.	Broadbelt, Winifred	Ministry of Transport, Public Works and Water Management / The Netherlands	wini.broadbelt@minienm.nl
9.	Brown, Christopher	International Seabed Authority / Jamaica	chrisgbrown@live.co.uk
10.	Buhmann, Sitta	Friends of the Earth Germany (BUND) / Germany	Sitta.buhmann@bund-bremen.net
11.	Busse, Lilian	German Environment Agency / Germany	lilian.busse@uba.de
12.	Christiansen, Sabine	Institute for Advanced Sustainability Studies / Germany	sabine.christiansen@iass-potsdam.de
13.	Clark, Malcolm	NIWA / New Zealand	malcolm.clark@niwa.co.nz
14.	Claussen, Ulrich	German Environment Agency / Germany	Ulrich.claussen@uba.de
15.	Cobley, Amber	National History Museum / UK	acc2g10@soton.ac.uk
16.	Cormier, Roland	Helmholtz-Zentrum Geesthacht, Centre for Materials and Coastal Research / Germany	roland.cormier@hzg.de
17.	Craik, A. Neil	School of Environment, Enterprise and Development at the University of Waterloo / Canada	ncraik@uwaterloo.ca
18.	Currie, Duncan	GLOBELAW / New Zealand	duncanc@globelaw.com
19.	Damian, Hans-Peter	German Environment Agency / Germany	hans-peter.damian@uba.de

Nr.	NAME	AFFILIATION	EMAIL
20.	Davis Mattis, Laleta	University of the West Indies / Jamaica	laletadavismattis@yahoo.com
2 .	De Wachter, Tom	DEME Group / Belgium	de.wachter.tom@deme-group.com
22.	Dom, Ann	Seas at Risk / Belgium	adom@seas-at-risk.org
2ɔ.	Durussel, Carole	Institute for Advanced Sustainability Studies / Germany	Carole.durussel@iass-potsdam.de
2ₐ.	Eder, Tim	Federal Ministry of Education and Research / Germany	tim.eder@bmbf.bund.de
25.	Ermakova, Livia	VNIIOkeangeologia / Russia	Livia77@inbox.ru
2€.	Flynn, Adrian	FATHOM PACIFIC / Australia	adrian.flynn@fathompacific.com
27.	Fuentes, Carlos Iván	United Nations, Division for Ocean Affairs and the Law of the Sea / USA	fuentesc@un.org
28.	Fukushima, Tomohiko	Japan Agency for Marine-Earth Science and Technology	fukushimat@jamstec.go.jp
29	Gianni, Matthew	Deep Sea Conservation Coalition / The Netherlands	matthewgianni@gmail.com
30	Ginzky, Harald	German Environment Agency / Germany	harald.ginzky@uba.de
31	Gjerde, Kristina	Policy Advisor for the IUCN Global Marine Program / Poland	kristina.gjerde@eip.com.pl
32.	Gollner, Sabine	German Centre for Marine Biodiversity Research / Germany	sabine.gollner@senckenberg.de
33.	Haeckel, Matthias	GEOMAR / Germany	mhaeckel@geomar.de
34.	Hamann, Kristin	GEOMAR / Germany	khamann@geomar.de
35.	Houghton, Katherine	Institute for Advanced Sustainability Studies / Germany	katherine.houghton@iass-potsdam.de
36.	Imhoff, Heike	Federal Ministry for the Environment, Nature Conservation, Building and Nuclear Safety / Germany	heike.imhoff@bmub.bund.de
37.	Jaeckel, Aline	Macquarie Law School at the Macquarie University /Australia	aline.jaeckel@mq.edu.au
38.	Janssen, Felix	Max-Planck-Institute / Germany	fjanssen@mpi-bremen.de
39.	Jenisch, Uwe	Deep Sea Mining Alliance / Germany	jenisch@deepsea-mining-alliance.com
40.	Jones, Daniel	National Oceanography Centre / UK	dj1@noc.ac.uk

Nr.	NAME	AFFILIATION	EMAIL
41.	Ju, Se-Jong	Korea Institute of Ocean Science & Technology	SJJU@KIOST.AC.KR
42.	Kanda, Naohisa	Japan NUS Co., Ltd.	Kanda-n@janus.co.jp
43.	Kaschinski, Kai	Fair Oceans / Germany	fair-oceans@gmx.info
44.	Knapp, Rodin	Federal Ministry for Economic Affairs and Energy / Germany	rodin.knapp@bmwi.bund.de
45.	Koyano, Mari	Faculty of Law at the Hokkaido University / Japan	koyano@juris.hokudai.ac.jp
46.	Krivanek, David	Federal Foreign Office / Germany	504-1@auswaertiges-amt.de
47.	Kühne, Hartmut	Federal Ministry for Economic Affairs and Energy / Germany	hartmut.kuehne@bmwi.bund.de
48.	Küper, Arne	Federal Ministry for Economic Affairs and Energy / Germany	buero-vb2@bmwi.bund.de
49.	Larsen, Petra	University of Manchester School of Law / UK	petra.larsen@manchester.ac.uk
50.	Laugesen, Jens	DNV GL group of companies / Norway	jens.laugesen@dnvgl.com
51.	Lee, Rena	Attorney-General's Chambers / Singapore	rena_lee@agc.gov.sg
52.	Levin, Lisa	Scripps Institution of Oceanography / USA	llevin@ucsd.edu
53.	Liu, Shaojun	Central South University / China	liushaojun@csu.edu.cn
54.	Lodge, Michael	International Seabed Authority / Jamaica	mlodge@isa.org.jm
55.	Machetanz, Kurt	State Authority for Mining, Energy and Geology / Germany	kurt.machetanz@lbeg.niedersachsen.de
56.	Maki, Janet	Janet G. Maki PC / Cook Islands	tagilaei@gmail.com
57.	McQuaid, Kirsty	Plymouth University / UK	Kirsty.mcquaid@plymouth.ac.uk
58.	Merck, Thomas	Federal Agency for Nature Conservation / Germany	thomas.merck@bfn.de
59.	Miller, Kathryn	Greenpeace Research Laboratories / UK	isunit@greenpeace.org
60.	Mißling, Sven	Project Management Jülich / Germany	s.missling@fz-juelich.de
61.	Müller, Christian	Federal Institute for Geosciences and Natural Resources / Germany	Christian.mueller@bgr.de

Nr.	NAME	AFFILIATION	EMAIL
62.	Narberhaus, Ingo	Federal Ministry for the Environment, Nature Conservation, Building and Nuclear Safety / Germany	ingo.narberhaus@bmub.bund.de
63.	Ndougsa Mbarga, Théophile	Ministry of Mines, Industry & Technological Development / Cameroon	theopndougsa@gmail.com
64.	Nugent, Conn	Pew Charitable Trusts / USA	cnugent@pewtrusts.org
65.	Oliveira, Carina	Law School at the University of Brasilia / Brazil	carina2318@gmail.com
66.	Paterson, Gordon	Natural History Museum, Department of Life Sciences / UK	g.paterson@nhm.ac.uk
67.	Perez, José Angel	UNIVALI – Universidade do Vale do Itajaí / Brazil	Angel.perez@univali.br
68.	Pingel, Jan	Association of Protestant Churches and Missions in Germany	jan.pingel@emw-d.de
69.	Ramirez-Llodra, Eva	Norwegian Institute for Water Research	eva.ramirez@niva.no
70.	Reichert, Christian	Federal Institute for Geosciences and Natural Resources / Germany	chrisj.reichert@t-online.de
71.	Reydellet, Bertrand	Sea Europe / Belgium	br@seaeurope.eu
72.	Ribeiro, Marta Chantal	Faculty of Law at the University of Porto / Portugal	mchantal@direito.up.pt
73.	Rodriguez, Donna M.	Minister and Consul of the Philippines	Donna.rodriguez@philippine-embassy.de
74.	Rong, Wang	Ocean Mineral Singapore Pte Ltd.	wang.rong@KOMtech.com.sg
75.	Rozemeijer, Marcel	Wageningen University & Research / The Netherlands	marcel.rozemeijer@wur.nl
76.	Rühlemann, Carsten	Federal Institute for Geosciences and Natural Resources / Germany	carsten.ruehlemann@bgr.de
77.	Sampat, Payal	Earthworks / USA	psampat@earthworksaction.org
78.	Sharma, Rahul	National Institute of Oceanography / India	rsharma@nio.org
79.	Smith, Samantha	Nauru Ocean Resources Inc. / Nauru	samantha@nauruoceanresources.com
80.	Spickermann, Ralph	Lookheed Martin Corporation / USA	ralph.spickermann@LMCO.com
81.	Steinbach, Volker	Federal Institute for Geosciences and Natural Resources / Germany	Volker.steinbach@bgr.de

Nr.	NAME	AFFILIATION	EMAIL
82.	Stoyanova, Valcana	Interoceanmetal Joint Organization / Poland	valcana.stoyanova@gmail.com
83.	Sugishima, Hideki	Japan Oil, Gas and Metals National Corporation	sugishima-hideki@jogmec.go.jp
84.	Tauchi, Tomoko	Deep Ocean Resources Development Co., Ltd. / Japan	tomoko@dord.co.jp
85.	Thiele, Torsten	Global Ocean Trust / UK	tors10th@me.com
86.	Unger, Sebastian	Institute for Advanced Sustainability Studies / Germany	sebastian.unger@iass-potsdam.de
87.	van Nijen, Kris	DEME Group / Belgium	van.nijen.kris@deme-group.com
88.	Ventura, Victor	Faculty of Law, University of Hamburg / Germany	vfventura@gmail.com
89.	Verlaan, Philomène	Dept. of Oceanography at the University of Hawai'i	pverlaan@gmail.com
90.	Villanueva, Maria-Sofia	European Commission / Belgium	Maria-Sofia.VILLANUEVA@ec.europa.eu
91.	Vink, Annemiek	Federal Institute for Geosciences and Natural Resources / Germany	annemiek.vink@bgr.de
92.	Walker, Michelle	Ministry of Foreign Affairs and Foreign Trade / Jamaica	michelle.walker@mfaft.gov.jm
93.	Walmsley, Simon	WWF UK, The Living Planet Centre	swalmsley@wwf.org.uk
94.	Warner, Robin	Australian National Centre for Ocean Resources and Security	rwarner@uow.edu.au
95.	Warren, Jennifer	Lookheed Martin Corporation / USA	jennifer.warren@LMCO.com
96.	Watzel, Ralph	Federal Institute for Geosciences and Natural Resources / Germany	gabriele.herbst@bgr.de
97.	Weaver, Philip	Seascpae consultants / UK	phil.weaver@seascapeconsultants.co.uk
98.	Xu, Lily Xiangxin	Kiel University / Germany	Xuxiangxin424@gmail.com
99.	Xu, Xuewei	Second Institute of Oceanography / China	xuxw@sio.org.cn
100.	Xue, Guifang	KoGuan Law School at the Shanghai Jiao Tong University / China	juliaxue@sjtu.edu.cn